T0342285

SCARCITY

SCARCITY

A History from the Origins of Capitalism
to the Climate Crisis

Fredrik Albritton Jonsson and Carl Wennerlind

HARVARD UNIVERSITY PRESS

Cambridge, Massachusetts & London, England | *2023*

First printing

Library of Congress Cataloging-in-Publication Data

Names: Jonsson, Fredrik Albritton, 1972–, author. | Wennerlind, Carl, author.
Title: Scarcity : A history from the origins of capitalism to the climate crisis /
    Fredrik Albritton Jonsson and Carl Wennerlind.
Description: Cambridge, Massachusetts : Harvard University Press, 2023. |
    Includes bibliographical references and index.
Identifiers: LCCN 2022037064 | ISBN 9780674987081 (cloth)
Subjects: LCSH: Scarcity. Economics—Philosophy. | Nature—Effect of human beings
    on. | Capitalism—Europe. | Europe—Economic policy. | Europe—Intellectual life.
Classification: LCC HB801 .J66 2023 | DDC 306.3/42094—dc23/eng/20221025
LC record available at https://lccn.loc.gov/2022037064

To our Mothers

CONTENTS

SCARCITY

INTRODUCTION

Beyond One Concept of Scarcity

Seen from outer space, the global economy is imperceptible. To the naked eye, even the most ambitious engineering project loses definition and dissolves into earth, oceans, and sky. The roads and railways, factories, and suburbs all seem to vanish without a trace. Only on the night side of the earth does the modern economy come into view: millions of lights joined together in a planetary luminescence.

Lights alone do not tell the whole story, of course. To take in the planetary impact of the global economy, other forms of observation are necessary. Hundreds of monitoring stations across the world now map the impact of economic growth on the carbon cycle. No physical trend of the last century has had a more profound effect than the accumulation of greenhouse gases. In the atmosphere, carbon dioxide forms a trace gas of miniscule proportions, yet this tiny chemical fluctuation turns out to have calamitous consequences for the climate system over time. Since the nineteenth century, greenhouse gas emissions from manufacturing and other energy-intensive sectors have begun to nudge the earth system toward a new state. Humanity has left the relatively stable climate of the Holocene epoch and entered a new stage in the history of the planet, provisionally named the Anthropocene.[1]

The cumulative effect of all our economic actions casts a shadow across the atmosphere, locking in heat in the biosphere and thereby raising the annual mean temperature of the planet. This is the consequence of a highly peculiar phenomenon: exponential economic growth. For 99.9993 percent of the time that *Homo sapiens* has lived on earth, there was no sustained economic growth at all. Only in the last two, maybe three, centuries has economic growth become a natural part of human life—a seemingly unequivocal good essential to the thriving of humanity. Present generations find it difficult even to conceive of the world without the concept of economic growth. Since the seventeenth century, scientist and engineers have become more and more confident in their ability to control the natural world. Yet this new power is terrifyingly partial and perhaps far more blind than we realize. While humans have learned to split the atom, manipulate the genome, and put people on the moon, they have also inadvertently produced pollution and biodiversity loss on a planetary scale. The seventeenth-century project to control nature has given rise to a series of nightmarish side effects that are now jeopardizing the very conditions that have enabled the emergence of complex societies. Global environmental change is putting the future of the human species at risk.[2]

To address these problems, capitalist societies have to change the way they interact with the planetary environment. We need to alter the way we think about the economy and nature, as well as the relationship between the two. For the better part of the last century, much of our approach has been grounded in modern neoclassical economics and its fundamental axiom of scarcity. Because human desire for consumption is assumed to be insatiable and nature is by definition finite, economists reason that all humans and firms are forced to make tradeoffs to maximize their happiness and profits. This means that, at any given moment, economic actors seek to make the most efficient use of natural resources and, over time, they strive to develop science and technology to engender as much economic growth as possible. If, in this process, natural resources start running low, economists predict that entrepreneurs aided by new science will respond to higher prices and develop substitutes. The conception of nature as scarce, yet capable of infinite improvement and infinite substitutability, has proven remarkably effective in promoting economic growth and ever-expanding consumption. Yet this conception of scarcity is also at the heart of the planetary crisis we now face.

For some time now, scientists have warned of sweeping, systemic changes to the earth system caused by fossil fuel economies and over-consumption. Anthropogenic climate change is the best-known threat. Greenhouse gases are pushing the planet toward new extremes of heat, humidity, drought, and flood. These changes will likely lead to a decline in agricultural productivity in key regions. Global warming will cause sea levels to rise, threatening densely populated coastal areas that are particularly vulnerable to rising seas. Oceanic ecosystems are also under increasing strain. Acidification threatens vital biota like coral reefs and phytoplankton. Closely linked to climate change is the trend of rising extinction rates. Climate deterioration and land-use change are rapidly reducing the terrestrial biodiversity that underpins the proper functioning of ecosystems and human economies. As if this were not bad enough, chemical pollution also poses unprecedented risks to the planetary environment and human well-being. Meanwhile, modern agriculture produces excess flows of nitrogen and phosphorus that damage the health of waterways and coastal ecosystems. Industrial agriculture and land clearance also appear to accelerate the emergence of new pathogens like COVID-19. With so many interrelated and escalating threats, capitalist societies appear to have reached a breaking point. Without fundamental transformation, humanity confronts planetary disaster.[3] We are therefore left with no other option than to reconsider fundamentally how we organize our economy.

To create an economy for the future, we need novel ways of thinking. To develop new ideas, we need to understand the past. This book condenses five hundred years of debates about the relationship between nature and economy, surveying how philosophers, political theorists, and economists in the past have conceived of this relationship. While historians often point out that knowing history prevents us from repeating it, we believe, more ambitiously, that historical knowledge not only allows us to avoid repetition but provides us with a shared understanding that can help us construct a better future. We hope that readers of this book, by gaining a better sense of how people in the past have conceived of the nature-economy nexus, will be inspired to think imaginatively about alternatives to the neoclassical idea of scarcity. We need to move toward an economy that is capable of meeting human needs at the same time that it allows for the earth system to operate in a manner that favors both human flourishing and the diversity of nonhuman life.

Although focused on the concept of scarcity, the centerpiece of modern economics, this book is not written from within the discourse of neoclassical economics. Rather, it locates economic thinking in a much broader historical context. We hope that many scholars, including anthropologists, historians, sociologists, political scientists, and economists, will find our historical approach useful. Our main purpose, however, is to reach concerned global citizens intent on pursuing solutions to the looming planetary crisis. Much of the argument here took shape in the classroom as we debated these ideas with undergraduates. We have written the book with students and other young people in mind, trying to make our ideas as accessible as possible, even to newcomers to intellectual history.

Varieties of Scarcity

The past is filled with different ways of thinking about scarcity. Since the sixteenth century, an array of philosophers, political theorists, economic theorists, even novelists and poets, have sought to identify and articulate the "ideal" relationship between nature and the economy. In their writings, we find a diverse set of ideas about scarcity: its sources, its implications, and its demands on human actors. The organization of this book reflects what we consider to be major trends and shifts in the evolution of these historical conceptions. Moving roughly chronologically through the centuries to our present day, each chapter identifies past ideas of scarcity that emerged in contemporary writings about the nature-economy relationship. To make sense of these different intellectual currents, we have grouped ideas that are closely related under a common rubric. Our names for these types of scarcity are not necessarily actor's categories—that is to say, they were not used by people in the past. Yet by categorizing and naming these past versions of scarcity, we have put together a rich and long-term history of a concept that today is all too often considered synonymous with neoclassical economics. The modern neoclassical conception of scarcity emerged only in the 1870s. Prior to that moment, people understood the nature-economy nexus in many different ways. We seek to make clear that scarcity itself can and should be liberated from its connotations in modern economics.

While this book names several distinct historical conceptions of scarcity, most ultimately fall within one of two umbrella categories. *Cornucopian* ideologies include a series of ideas that endorse an active mastery of nature together with a dynamic and expansive notion of desire. All versions of Cornucopianism are rooted in optimism that nature's resources, however limited, can be extended infinitely by humans—although as we will see, they often differ on how exactly to improve nature's bounty and how expansively to embrace human desires. The category of Cornucopian ideologies includes what we call Cornucopian Scarcity, Enclosure Scarcity, Enlightened Scarcity, Capitalist Scarcity, and Neoclassical Scarcity. This tradition first emerged in the seventeenth century and eventually reached a dominant position by the end of the nineteenth century. As an intellectual current, Cornucopianism has helped push us headlong down our current path of ever-expanding economic growth and planetary crises.

The category of *Finitarian* scarcity, meanwhile, emphasizes the limits to human power over nature and the need for constraint and moderation of human desires. As we shall see, the Finitarian ideologies featured in this book variously perceive the reasons for these limits and offer different approaches to constraining human desires. But at their base, these ideologies are rooted in the fundamental belief that nature's abundance is finite—and that human desires must be curbed to maintain a balance between nature and economy. The category of Finitarian ideologies consists of what we term Neo-Aristotelian Scarcity, Utopian Scarcity, Malthusian Scarcity, Romantic Scarcity, Socialist Scarcity, and Planetary Scarcity. We note that Enclosure Scarcity and Socialist Scarcity can be understood as composites of Finitarian and Cornucopian forms.

Although the Romantic and Socialist versions of scarcity have had a powerful and recurring influence over culture and politics, only their sixteenth-century predecessor, Neo-Aristotelian Scarcity, achieved cultural hegemony in the West. Finitarianism therefore primarily represents a history of resistance and aspiration rather than dominance. Yet we would be remiss to underestimate Finitarianism's intellectual force, which drives not only its ability to attract devoted adherents, but also its power to stimulate creative thinking about alternative futures.

Finitarian and Cornucopian worldviews developed side by side in mutual opposition. Conflict bonded them together, such that each side defined itself by rejecting the other. Because they sought to answer the same question—are there limits to economic growth?—they often ended up feeding on each other, generating rival forecasts of the future and competing conceptions of the public good. The intellectual currents we examine in this book demonstrate how Finitarianism and Cornucopianism emerged as oppositional intellectual frameworks. We might think of their development as a form of family feud inherited from one generation to another, always locked in battle, but producing new grievances and new areas of conflict over time. The fear of limits to economic growth provoked optimistic visions of abundance, which in turn came under attack by critics. Of course, this conflict did not happen in a material vacuum. Both sides looked to the natural world and technology to justify their positions: where Cornucopians celebrated the bounty of natural resources, the power of human ingenuity, and the insatiability of desires, Finitarians emphasized limits, unintended consequences, and simple needs.

Nearly all of the ideas of scarcity that we examine in this book are part of the system of *capitalism*. Capitalism, although difficult to define as it assumes so many different forms across time and space, we take to be a social system that emerged for the first time in Europe during the early modern era (circa 1500–1800). The capitalist order is based on the institutions of private property, markets, money, profits, capital, corporations, and wage labor. Some of these institutions can be found in earlier social systems, but when we add that, in capitalism, competition, entrepreneurship, consumerism, colonization, commodification, specialization, and scientific progress serve the larger purpose of capital accumulation, we inch closer to a robust definition. We also need to include a centralized state that is capable of intervening, regulating, and legislating in a manner that promotes the expansion and stability of capitalism. All of these characteristics do not have to be simultaneously present for us to view a society as capitalist—after all, capitalism contains both free and enslaved labor, free competition and monopolies, private and public property, free trade and protectionism, democratic and authoritarian states. Yet, the fewer of these institutions a society contains, the further away from capitalism it drifts. We can also define capitalism by looking for its ecological footprint. Capitalist accumulation requires intensifying exploitation of the local and global environment through processes of commodification, extraction, and scientific management. Lastly, we need to take seriously capitalism's capacious ideological apparatus, with room for numerous conflicting ideologies. Without able intellectual defenders, capitalism could never have become a dominant social and political force in the world.

Five Hundred Years of Scarcity

To understand how the rivalry between Cornucopian and Finitarian forms of scarcity emerged, we must begin by considering the notions of limits and abundance in preindustrial societies. Prior to the age of capitalism, the nature-economy nexus was conceived of in a variety of ways. Anthropologists have found evidence of hunter-gatherers enjoying "affluence without abundance."[4] Paleolithic foragers did not have much, but because their wants were small, they always had more than they needed. Only in the aftermath of the Neolithic Revolution, when new institutions emerged based on centralized power and sedentary populations, did the view of nature and desire shift. The formation of agricultural societies was made possible by the interglacial epoch known as the Holocene which began 11,700 years ago. While the early Holocene was considerably warmer than the last few millennia, the trend overall was toward relative stability. Carbon dioxide levels in the atmosphere during the Holocene varied between 260 and 285 parts per million while the temperature shifted only very little, about one degree Celsius up or down from the global average. Internal variations like the Roman and Medieval Warm Period or the drop in temperatures during the seventeenth-century Little Ice Age were trifling compared to the great cycles of the Pleistocene. This relative stability of climate allowed agricultural societies to rely on predictably recurring cycles and flows within the organic economy.[5]

In agrarian societies, people began to conceive of the social order as a steady oscillation between physical scarcity and material plenty. The biblical notion of seven good years followed by seven years of famine captured the prevailing fatalism. The word *scarsete* or *skarcete* first appeared in Middle English during the fourteenth century as a loan from the Old French *escharseté*.[6] During this period scarcity referred specifically to the insufficient supply of necessities to feed the common people. It was an earthly phenomenon, produced by bad weather and harvest failures. When dearth proved persistent, it led to subsistence crisis and mass death, unless societies maintained emergency supplies.

Even during years of relatively abundant harvests, there was a general sense of finitude. Along with these material constraints, a moral imperative to curb human appetites also emerged. According to the Christian worldview of the sixteenth century, as we show in Chapter 1, the relationship between human desires and nature was conceived as a delicate balance of limitations. Religious doctrine made it clear that pious people never let their desire for pleasure, of any kind, run amok. When kept within socially and spiritually circumscribed limits, desires could exist in harmony with nature's limited yield. People were expected to respect the inherent restrictions of nature and make do with the little they had. It is thus in sixteenth-century Europe that we locate the earliest Finitarian model, and the only one to achieve any kind of cultural hegemony: Neo-Aristotelian Scarcity. Losing control over one's desires was, as Aristotle had said long ago, tantamount to losing one's humanity. Yet even in the 1500s, these ideas about human desire were challenged by a growing culture of commerce and enclosure that spurred critiques and alternatives from the likes of Thomas More and Martin Luther.

A radically new Cornucopian way of conceiving of the relationship between nature and the economy emerged in the seventeenth century, starting the slow and circuitous route toward the modern neoclassical concept of scarcity. As Chapter 2 shows, the natural philosopher and politician Francis Bacon popularized the idea that humanity could, with the aid of scientific knowledge, bring nature under control and force it to share its dormant riches. Bacon's disciple, Samuel Hartlib, praised nature as an infinite treasure, capable of giving rise to earthly abundance. Soon thereafter, the economic writer, fire-insurance entrepreneur, and London real estate tycoon Nicholas Barbon endorsed insatiable desires as not only natural, but also socially beneficial. In contrast with the traditional notion of harmonious limitations, scarcity was now seen as the product of intertwining infinities: the endless human desire for consumption and infinitely expandable nature. We describe these ideas collectively as Cornucopian Scarcity, reflecting their position as progenitors of later Cornucopian ideologies that developed across the ensuing centuries. Unlike its sixteenth-century predecessors, Cornucopian Scarcity legitimized boundless wants as the force that—supported by scientific advances—would propel the infinite improvement of nature and hence infinite human progress.

Paradoxically, Bacon and Hartlib's dream of godlike power took shape in the midst of the Little Ice Age, when mean temperatures in Europe decreased by one degree Celsius. While we do not have a full picture of how climate deterioration challenged seventeenth-century society, we know that Hartlib and his circle sprang to action during the harsh winters and near-famine conditions of the 1640s and 1650s. Like most people in the period, they regarded the relationship between the economic order and the climate as interdependent. If the landscape could be brought under scientific control, "savage" nature would become "civilized."[7] Writing in the following century, the Scottish Enlightenment philosopher David Hume explained, according to the logic of the day, that the warming trend was the result of the fact that "the land is at present much better cultivated, and that the woods are cleared, which formerly threw a shade upon the earth, and kept the rays of the sun from penetrating to it."[8]

If the seventeenth century witnessed a radically new form of thinking about nature and economy, the eighteenth century—the focus of Chapter 3—emphasized gradual progress. While much of the previous century's optimism survived, Enlightenment-era thinkers were not quite as enamored with the ideas of infinite human desire and endlessly bountiful nature. David Hume, Daniel Defoe, and Adam Smith, among others, argued that nature could provide great—but perhaps not endless—wealth. For example, Hume suggested that nature was always scarce but that it was possible, through industriousness and scientific progress, to slowly extend its boundaries. As long as human creativity remained vibrant, there were no absolute limits to growth. Enlightenment-era thinkers were also more inclined to believe that humans, while drawn to consumption, should temper their selfish desires. Hume argued that commercial civilization, political liberty, and liberal education would refine and redirect human desires onto a higher plane. Civilized people would become more prone to poetry and philosophy than to rampant consumption of luxuries. The Enlightenment version of scarcity therefore envisioned an incremental curtailment of initially strong desires for material affluence and a gradual, scientifically engineered, expansion of nature's bounty. This was a more sensible and measured form of Cornucopianism, in which the future held the promise of a partial easing of the yoke of scarcity.

While the Hartlibians and Hume opened a path for the modern notion of scarcity, there were contemporary voices who objected to the ideology of infinite growth and infinite consumption. During the seventeenth century, the anti-Enclosure militant Gerrard Winstanley put forth a radical critique of property and money, arguing that they polarized society and drove a wedge between rich and poor. Winstanley saw scarcity as a universal condition, experienced even during moments of abundant harvests and economic flourishing, with the rich constantly striving for more and the poor always fearing starvation—a condition we will call Enclosure Scarcity. Winstanley was responding to the violence of the enclosures— the first crucial step in agrarian capitalism whereby the land was transformed from a shared space of common use and existential meaning to an economic resource accumulated in the hands of the few. About a century later, a philosophical contrarian from Geneva, Jean-Jacques Rousseau, argued that the constant desire for consumption had contaminated the social fabric. Everything in society and nature had become subjugated to the quest for trivial luxuries, resulting in the corruption of the good life. Insatiable desires, infinite growth, and perpetual scarcity, Winstanley and Rousseau argued, were social constructs that had a beginning and should have an end. They each formulated their own Finitarian vision of the world.

The full flourishing of the Industrial Revolution in the nineteenth century fostered a new wave of Finitarian notions of scarcity. Chapter 4 examines the Romantic writers who launched a systematic rethinking of both human desire and nature. These thinkers imagined a world in which people were motivated by beauty and community rather than consumption, and treated nature as the spiritual center of human life. What we call Romantic Scarcity embraced human restraint and material simplicity that respected the finite resources and the transcendental value of the natural world. At the same time, Thomas Robert Malthus pessimistically argued that the needs of a geometrically growing population would soon outstrip the agricultural yield, since the latter could grow only at an arithmetic rate. Disease, war, and famine would cull the surplus population until, after much suffering, the excess numbers were brought back into balance with a strictly circumscribed natural world. As Chapter 5 shows, Malthusian Scarcity held that the world's finite supply of land placed immovable physical limitations on human growth.

The revolutionary changes underway in the nineteenth century also sparked Karl Marx's radical critique of both the Enlightenment and Malthusian versions of scarcity. Marx argued that scarcity was driven not by the boundless desire for consumption but rather by industrial capitalists' incessant pursuit of capital accumulation. He shifted the blame for scarcity from humanity in general to the emerging industrial capitalist class, who constantly sought to impose labor on the working classes not primarily to enjoy the fruits thereof but to reproduce their dominance. Since the incessant force driving the capitalists, as a class, was the reproduction of command and control, we call it Capitalist Scarcity. Together with other radicals, such as Robert Owen and Charles Fourier, Marx envisioned an alternative future, one based on an entirely different relationship between nature and the economy. They looked forward to a world in which technology would produce an abundance of material wealth to satisfy all basic needs, while liberation from capitalist domination would free people to pursue the full spectrum of human passions, not just those that could be satisfied through consumption. Whereas Malthus identified the sources of scarcity in the clash between the earth's physical limits and the insatiability of collective human desire, the originators of what we call Socialist Scarcity saw the future as the interplay between a needs-based economy and the scientifically driven mastery of nature.

After the disruption of the Little Ice Age, the climate of the northern hemisphere grew more favorable during the Enlightenment and the nineteenth century. This warm spell coincided with the wide-ranging adoption of fossil fuel, first in Britain and then across the West. A geological endowment, stored up over millions of years, enabled a quantum leap in energy use during the nineteenth and twentieth centuries. Yet precisely this windfall also disrupted the carbon cycle that controls the planetary climate. By unleashing carbon dioxide into the atmosphere on a scale never seen before, the new fossil fuel economy brought the Holocene epoch to an end. Carbon dioxide levels indicate clearly that this shift had happened already by the end of the nineteenth century, just as socialist theorists and marginalist economists launched their rival bids to remake the world.[9]

Marx's vision of an overthrow of existing social relations shook the European bourgeoisie to the core. Liberal thinkers set out to develop an alternative ideology, one that put capitalism in a more favorable light.

Nearly simultaneously, William Stanley Jevons, Léon Walras, and Carl Menger developed what would become modern neoclassical economics. The version of scarcity at the heart of the new economic discourse that they pioneered had very little to do with either the problem of poverty or the challenge of resource exhaustion. Instead, their version of scarcity, explored in Chapter 7, was a philosophical conjecture that originated in the assumption of insatiable human wants and infinite substitutability on the one hand, and on the other hand, the fact that all resources are by definition finite. They all argued that, while people who experienced poverty or confronted dwindling natural resources certainly faced scarcity, their experiences differed only by degree from those of everyone else. Without tapping into the scholarship in anthropology or psychology, economists alleged that people everywhere confronted the same universal condition of scarcity. This textbook example makes their position clear: "Small bands of African Bushmen face it; so do Amazon Indians and Greenland Eskimos. Peasants in China, Egypt, and Peru suffer from it; so do urban dwellers in Moscow, Paris, and New York. All of them, every day, wrestle with the basic economic problem of scarcity."[10] To be human thus means to be involved in the Sisyphean task of constantly striving for abundance in the context of inescapable scarcity. Regardless of how much wealth is attained or how it is distributed, the nagging desire for more never goes away. This version of Cornucopianism was systematized and popularized by the neoclassical economists, starting with the London School of Economics professor Lionel Robbins. "We have been turned out of Paradise," he began. "We have neither eternal life nor unlimited means of gratification. Everywhere we turn, if we choose one thing we must relinquish others." Robbins concluded: "Scarcity of means to satisfy ends of varying importance is an almost ubiquitous condition of human behavior."[11] It is one of history's many ironies that at the same time that the West enjoyed a golden age of unprecedented affluence (1945–1975), scarcity became the centerpiece of economic analysis.

In the twentieth century, fossil fuel–induced economic growth gathered further momentum as petroleum and natural gas facilitated the development of new technologies, from international air travel to synthetic fertilizer. The sustained boom after World War II led to an escalation of carbon emissions, increasing the level of atmospheric carbon dioxide from 311 parts per million (ppm) in 1950 to 331 ppm in 1975. Cheap energy ushered in unprecedented affluence in the advanced economies of the world, but also set the planet on the path toward multiple tipping points. By the end of the twentieth century, the new, interdisciplinary field of earth system science illuminated the risks created by runaway growth to the stability of the system. The discovery of ozone depletion in the 1980s brought home to a stunned world how seemingly trivial forms of consumption could lead to planetary danger. Common household goods like refrigerator coolants and shaving cream posed a lethal threat to the safety of the biosphere. Around the same time, anthropogenic climate change entered into public awareness. More and more voices warned that the old dream of godlike power over nature had opened a Pandora's box of environmental horrors.

By the early twenty-first century, the dominant idea of Neoclassical Scarcity was on a collision course with a new understanding of the world: Planetary Scarcity, which we take up in Chapter 8. In 2000, the atmospheric chemist Paul Crutzen and the ecologist Eugene Stoermer coined the term Anthropocene to draw attention to the dramatic rupture in the history of the planet. Rapid economic growth based on fossil fuel use had forced the earth out of the Holocene and into a new geological epoch. From the beginning, the concept of the Anthropocene included a host of threats besides climate change. The Planetary Boundaries framework, devised by the environmental scientist Johan Rockström, described nine major tipping points that had the capacity to force the earth out of its Holocene state: climate change, biosphere integrity, land use change, freshwater use, biochemical flows, ocean acidification, atmospheric aerosol loading, stratospheric ozone deletion, and novel chemical entities. These nine boundaries revealed a tragic flaw in the Cornucopian conception of scarcity embraced in mainstream economics. Instead of seeing the natural world as a boundless stock of resources to control and command, earth system science models suggested that exponential economic growth was producing more pollution than the planet could absorb, risking major disruption to the safe functioning of the system.[12]

The growing threat to the global environment served up a frightening twist on the old fear of natural limits to growth, expanding the problem of finite *stock* to a scarcity of *sinks*. Energy and matter flow through the earth system between different reservoirs. When the flux of matter into a reservoir is greater than the outflow, the reservoir is defined as a

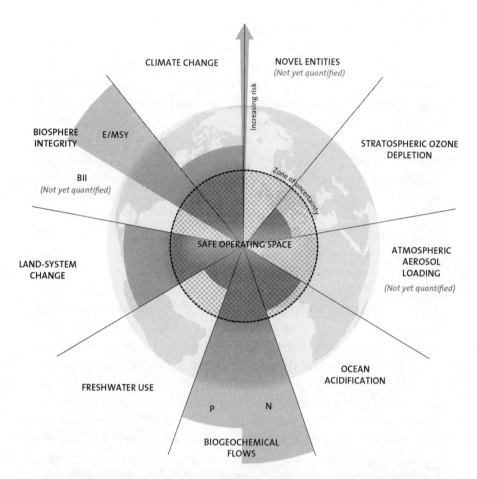

Planetary Boundaries, 2015. The Planetary Boundaries model defines the human economy as a subset of the global environment. Each of the nine boundaries suggests a quantitative measure for safe development. *Credit:* Stockholm Resilience Center.

sink. When coal, oil, and natural gas are burned, carbon dioxide is transferred from the ground to the atmosphere. Naturally occurring processes gradually remove carbon from the atmosphere and sequester it in sinks like the oceans, terrestrial vegetation, rocks, and soil, but the capacity of these sinks to store carbon dioxide is not unlimited. Beyond a certain threshold, excess carbon in the atmosphere will trigger a cascade of tipping points that undermine the safe functioning of the system.[13]

At the same time, earth system science also pointed to a second closely related planetary crisis of biodiversity. Rapid land use change and climate change threatened to unleash a sixth mass extinction in the near future. Here, too, science challenged the idea of nature as a mere stock of resources for human use. By defining biodiversity as a nonrenewable and irreplaceable foundation for all life, ecologists insisted that there were sharp limits to human power over the earth. These warnings have only grown louder in recent years. The Finitarian concept of Planetary Scarcity captures this tension, acknowledging that the earth system itself can and will be overwhelmed by insatiable wants and endless growth.[14]

Under these manifold pressures, neoclassical economics came under attack from a variety of directions. Many of its most prominent advocates sought to address these critiques by revising the neoclassical doctrine. Just to mention a few, in the 1970s the Hungarian émigré Tibor Scitovsky and the American economist Richard Easterlin revised the more-is-better assumption.[15] Around the same moment, the Oxford-trained economist Fred Hirsch argued against the idea that economic growth necessarily contributes to the quality of life.[16] Harvard-economist Amartya Sen launched a new form of development economics centered on freedom and quality of life rather than the "narrower views" fixated on gross national product or industrialization.[17] More recently, the British economist Karen Raworth fused the Planetary Boundaries framework with a universal model of social and economic development. She, too, rejected the growth ethos of conventional economics in favor of satisfying all humans' basic needs within the ecological limits set by Planetary Boundaries. She accused the neoclassical economists of adopting a model of scarcity that neglected the moral ends and environmental constraints of actual economic life.[18] Finally, Cambridge economist Partha Dasgupta has developed a program for the economics of biodiversity, focusing on not just physical capital but also human capital and natural capital.[19] Many other efforts are currently underway within economics to address previous shortcomings. Yet the extent to which academic economists have reoriented their research agendas should not be overstated. Out of nearly nineteen thousand articles in the top five economics journals between 1957 and 2019, "climate change" and "global warming" appeared only twenty-six times in the titles and thirty-two times in the abstracts.[20] Moreover, most of the economics curriculum taught at universities around the world remains faithful to its traditional principles. As such, the conception of scarcity that informs how policy makers, journalists, and business leaders approach the world is still very much grounded in the canonical version of neoclassical economics. To be absolutely clear, this book does not offer a critique of the usefulness or instrumentality of the neoclassical concept of scarcity—instead, the problem we highlight is that it has been *far too* successful. That is, by promoting the optimal use of resources and maximum economic growth, it has fostered a world in which the economy and nature are on a collision course. The primary aim of this book is therefore to expand our intellectual toolbox by drawing on how people in the past have understood the sources, meanings, and repercussions of scarcity, so that we can transcend the current hegemony of neoclassical economics.

The Power of Ideas

As intellectual historians, we believe that the manner in which people make sense of the world deeply shapes their actions. Each generation produces a world in the image of its ideas. The institutions we form, the policies we implement, the laws we pass, and the practices we pursue are undeniably structured by the prevailing worldview. As the great economist John Maynard Keynes declared:

> The ideas of economists and political philosophers, both when they are right and when they are wrong, are more powerful than is commonly understood. Indeed the world is ruled by little else.[21]

This is not to suggest that ideologies govern everything or that history unfolds according to a simple inherent logic, only that ideologies play a profoundly important role in shaping political agendas, legal changes, economic processes, and individual behavior. We reject deterministic models that see a one-to-one relationship between any particular society and its ideas; all societies are capable of producing an array of ideologies. While ideas always mirror the social structure, economic conditions, and political realities of their time, they also have the power to reshape these conditions to a significant degree.

The concept of ideology often has negative associations; it is seen as the opposite of the actual, the real, or the true. We employ the term differently here. For us, an ideology constitutes a worldview: a basic understanding of a society and how its constitutive parts fit together and acquire a discernable logic and purpose. An ideology offers a coherent perspective on a society that unifies its believers and creates a shared identity. Ideologies tend to be both rationalizing and legitimizing, in the sense that they provide "plausible explanations and justifications for social behavior which might otherwise be the object of criticism."[22] Ideologies can be said to be naturalizing, in the sense that they are often presented as natural, self-evident, and commonsensical. They are also frequently made to appear ahistorical, having no discernible beginning and thus no inevitable end. While ideologies pretend to be universal, applicable to everyone, they never achieve absolute dominance. Alternative ways of understanding the world are always available for those who seek them out.

Currently, the ideology of modern economics holds a powerful sway over the world. Neoclassical economists offer a coherent explanation of economic phenomena, and in so doing powerfully legitimize and encourage the maximization of efficiency, profits, utility, and growth. The theory also sets boundaries for what is considered real and common sense. Once students of economics accept the neoclassical notion of scarcity, only a particular understanding of the present and a limited set of future worlds become possible. Even though the actual conditions of modern capitalism do not look much like the models employed in modern economic theory, the theory nevertheless plays a critical role in structuring the modern understanding of capitalism. Thinking of nature as a storehouse of appropriable and tradeable material wealth alters how people interact with the earth system and all its elements. It makes it reasonable to conceive of the biosphere, first and foremost, as a standing reserve and a factor of economic growth.

If economists and politicians continue to use the modern neoclassical concept of scarcity to address the looming planetary crisis, they run the risk of generating solutions that only exacerbate the problems. They trap us in an intellectual framework that is unlikely to yield the kind of creative thinking we need. It certainly would be convenient if we were able to continue addressing our environmental problems with ever more economic growth—something humans have become very skilled at doing—but that is no longer an option. Yet the very idea of stepping off the infinite growth trajectory invokes multiple anxieties: we can no

longer be confident that each generation will be better off materially than
the previous one, modern pension systems might not remain solvent, and
we might not be able to generate enough jobs for everyone. Voluntarily
extracting ourselves from the infinite growth paradigm will require a
fundamental transformation in the way we think about and approach
the world.

Because this book examines ideas that shaped capitalism and modernity, we focus our attention on the writers in the Western canon.[23] This
means that we explore, for the most part, a narrow range of elite, white,
male thinkers, who enjoyed the privilege to publish their ideas and had access to institutions of intellectual authority. Among them are philosophers,
political, and social theorists, as well as economists, including Francis
Bacon, David Hume, Adam Smith, Carl Menger, and Alfred Marshall. Some
of these figures advanced ideas of infinite economic growth, in which we
can identify the roots of today's Neoclassical Scarcity. Others, however,
voiced oppositional discourses, among them Gerrard Winstanley, Jean-
Jacques Rousseau, Dorothy Wordsworth, Karl Marx, and Hannah Arendt.

For some of our readers, this book contains too many intellectual
figures already, whereas for others we have not included enough, omitting
writers who perhaps deserve to be included. We have aimed, however, to
select the thinkers who present each version of scarcity we want to highlight with the greatest lucidity. Had we tried to capture every nuance of
this genealogy, our book would have been many thousand pages long. We
have also aimed throughout the text to capture the ideas of those who
first formulated them, not to elaborate on subsequent debates and interpretations by our fellow academics. For this reason, endnotes are kept
to a minimum.

Resistance and opposition are, of course, not a monopoly of Western
dissidents. We recognize that there have been many other oppositional
voices, both within and outside the Western canon—voices that sang out
as capitalism spread around the globe. We do not believe that the Western canon should be the only fountain of ideas from which to draw when
thinking about the future, and we are strongly in favor of movements
toward a new global intellectual history. But we also recognize that other
scholars are better equipped to write this more expansive history of opposition to capitalism beyond the Western canon. Indeed, there is a rich and
growing scholarly literature on conceptions of the relationship between

nature and economy in subaltern and non-Western ideologies. Our hope is that readers ultimately will consider our book within the context of this larger global discourse, as part of an urgent, collective search for new paths of flourishing on the planet.[24]

While this book largely examines the past, our ultimate aim is to foster discussion about how to conceptualize the relationship between nature and economy in the future. It is still possible to change human behavior in ways that will enable societies to stay clear of the worst effects of the looming climate and biodiversity crises. This will require, however, a decisive shift beyond the hegemonic neoclassical conception of scarcity. It is in this sense that we believe the ideologies explored in this book can bear productively on our current planetary crisis. First, only by examining the development of ideas over centuries can we come to appreciate fully when and why perceptions of the relationship between nature and the economy changed. Second, a broader understanding of past ideas about scarcity gives us a comparative framework within which to evaluate what is historically specific about each concept and therefore how they differ. Third, and perhaps most importantly, the historical record shows that many different versions of scarcity have existed over centuries. The modern neoclassical version was never inevitable, but just one of many ideas of scarcity. If it was created at some point in time, it can also have an end: there is no universal truth or permanence to it. Finally, the historical record also offers a reservoir of alternative ways of thinking about nature and economy. None of them, of course, can be fully retrieved and reinstated in their original form, and some indeed are best left in the past. Others, however, might inspire us to think creatively about our future. In fact, as explored in the Conclusion, ongoing responses to Planetary Scarcity already contain some echoes of oppositional discourses from centuries past. Ideas do not neatly come and go, but tend to linger. Although each chapter of this book describes the emergence of a new concept of scarcity, none of these worldviews completely replaced the preceding ones or was replaced by those that followed. At any given point, there are numerous competing worldviews, old and new, vying for attention. One or a few ideologies might gain ascendancy for some time, but their triumph is never absolute. Nestled in the social fabric, they reappear in the future in a slightly different guise, once again ready to shape the course of history.

No matter which intellectual traditions frame our imagination of the future, one thing is clear: all societies face the same universal emergency in the twin threats of anthropogenic climate change and mass extinction. For a problem of this magnitude, there are no simple technical fixes. While innovation in technology and infrastructure will no doubt be indispensable, the deeper challenge we confront is how to rethink the relation between economy and nature. Instead of seeing economic activity as an independent power that masters nature, those of us who live in capitalist societies need new ways to understand production as a joint endeavor between humanity and earthly forces, from the microcosm of soil bacteria to the carbon cycle of the earth system. Instead of thinking of creativity as a purely human phenomenon, we should recognize how the natural world makes possible and shapes human agency and well-being. We also need to reorient the public debate toward new normative aims. After so much thoughtless destruction and degradation, ecological repair and restoration must become central priorities. At the same time, the work of repair must go hand in hand with new ideals of justice that help us overcome long-standing inequalities, within and between nation-states and continents.[25]

As historians we do not claim to have a ready and easy solution to any of these problems. But we insist that the reconstruction of the economic imagination will require historical detective work. We can only hope to free ourselves from the force of destructive ideas by understanding their historical roots. By delving into the past, we begin to see scarcity as historically contingent and tied to peculiar social and political contexts. Such inquiries widen the scope of our creativity in this moment of planetary emergency, clearing a space for new thought and action.

1

TYPES OF SCARCITY
BEFORE 1600

From the Book of Genesis, the faithful of the Judeo-Christian tradition have learned that God created the world for the benefit of humanity. Everything within nature—plants, flowers, trees, water, soil, minerals, fish, livestock, and wild animals—existed for the purpose of serving human needs. At first, in the Garden of Eden, Adam and Eve enjoyed the extraordinary beauties and bounties of Creation. The fruits of nature were available in abundance, enabling them to live blissfully without toil. After Adam and Eve ate from the Tree of Knowledge, however, God punished them by hiding nature's resources, making the ground hard and full of thorns and thistles. Humans now had to earn their food by the sweat of their brow. Nature still existed for the benefit of humanity, but it required people to spend the lion's share of their earthly days procuring food. As long as humans diligently worked the soil, nature would reward them with enough material wealth to get by. Yet no amount of toil could free them from the curse of Adam's act of disobedience against God. Human nature was permanently tainted by original sin. The physical world, too, was cursed as a place of tribulation, testing the faith of the believers with divine punishment: famine, earthquakes, floods, droughts, and plagues.

This Christian understanding of nature survived largely intact throughout Europe into the early modern period. The satisfaction of basic needs required a constant struggle with the material world. Hard work and prayers to God and the saints were seen as the only pathways to comfort and sufficiency. The Christian doctrine of a fallen world reflected the ecological constraints of agrarian societies. Photosynthesis set firm limits to economic expansion and population growth. Life was sustained almost entirely by solar energy, converted into grain, grasslands, and woodlands, which nourished people and animals, and into wind and water energy, which augmented muscular energy in the process of production. Manufactures, long-distance trade, and infrastructural improvements like canals and irrigation systems all boosted population and productivity, yet could not overcome ecological constraints. Coal—highly concentrated stores of solar energy—had been used to heat houses and melt metals in Europe for hundreds of years, but always on a miniscule scale.

While agrarian societies faced firm constraints, they also enjoyed certain advantages. Most important of these was the relative stability of the Holocene climate. In comparison with the volatile Pleistocene epoch, the natural world of the Holocene was far more benign. It is hardly a coincidence that agriculture emerged more or less simultaneously in many different places around the world by the end of the last Ice Age, some 11,700 years ago. The favorable climate was not just a material advantage but perhaps also a cognitive necessity. Only in a relatively stable and therefore *predictable* environment could complex, sedentary societies begin to flourish. Such long-term stability did not, however, preclude severe internal variations. In the early modern period, the global mean temperature fell by about one degree Celsius. While contemporaries were only dimly aware of this shift, latter-day climatologists have reconstructed this planetary trend and baptized it the Little Ice Age (the coldest stretch of which occurred between 1560 and 1725).[1]

Humanity's power to master nature was thus highly circumscribed. While life may not always have been, in the famous words of English philosopher Thomas Hobbes, "poor, nasty, brutish and short," people certainly lived in constant fear of bad harvests and food shortages.[2] Hunger was the "specter" haunting "early modern Europe, and one of the principal factors contributing to the insecurity of the age."[3] Getting food of sufficient quantity and quality was therefore the central preoccupation of people's lives; their physical health, mental well-being, and longevity depended on it. Communities were involved in a perennial struggle to maintain the proper balance between land, hands, and mouths. While all three of these factors were subject to change, they were more susceptible to negative shocks than sudden improvements. Because advancements in technology and techniques were yet infrequent, sustainable increases in the agricultural yield were rare. There were, unfortunately, relatively frequent sudden drops in agrarian output, resulting from bad harvests, pestilence, and war. In England during the sixteenth century, for example, approximately a quarter of harvests were deficient, with crisis conditions erupting during five periods: 1549–1551, 1555–1557, 1570–1574, 1586–1588, and 1594–1598. During these moments of impasse, people resorted to emergency measures such as baking bread with bran, acorns, sawdust, or hay, and boiling soups of water, roots, and grass with pieces of leather—perhaps a belt or shoe—to add a subtle flavor of meat. Malnourishment increased people's susceptibility to disease and illness, leading to moments of abrupt demographic decline. Stories of famines were common across early modern Europe. In Venice, a witness of the famine of 1577 described a harrowing scene in which "every evening in the Piazza San Marco, in the streets and at Rialto stand children crying 'bread, bread, I am dying of hunger and cold.'" Some fifty years later, a physician witnessing a famine in the Italian town of Bergamo reported that "most of these poor wretches are blackened, parched, emaciated, weak and sickly . . . they wander about the city and then fall dead one by one in the streets and piazzas and by the Palazzo."[4]

To make sense of humanity's place in and interaction with the natural world, early modern theologians, philosophers, political, and economic thinkers theorized the nature-economy nexus as a carefully calibrated balance of limitations. Although they did not use the term, we call this condition Neo-Aristotelian Scarcity, as it was inspired by the early modern interpretation of the great ancient philosopher. This worldview was centered on an understanding of nature as powerful yet limited and inflexible, and the economy as dedicated to subsistence and needs fulfillment rather than ever-expanding individual wants. Coinciding with the Reformation and the further development of a mercantile economy at the turn of the sixteenth century, a number of prominent intellectuals, including the German Martin Luther and the Englishman Thomas More, brought attention to the growing fissures between the Neo-Aristotelian notion of scarcity and the actualities of the commercial world. They argued that the greed unleashed in commercial societies undermined the restraints on desires and thus ruptured the delicate balance between the economy and nature. While Luther believed that the conditions of Neo-Aristotelian Scarcity could be restored, More insisted that the time for that idea had long passed. In his mind, a new and dreadful condition had already been established. Property and money had created incentives for the rich to accumulate endlessly, while the poor, who had been divorced from the land by the enclosures, lived in a constant state of deprivation. Because the enclosures generated a condition in which everyone always wanted more, we call this type of scarcity Enclosure Scarcity. To relieve this sorry state of human affairs, More submitted an ambitious vision for an alternative understanding of nature and economy, which we will call Utopian Scarcity.

These three early modern conceptualizations of scarcity—Neo-Aristotelian Scarcity, Enclosure Scarcity, and Utopian Scarcity—form the centerpiece of this chapter. None of these terms are actor's categories, in the sense that contemporaries used them, but they reflect the spirit and context of people's ways of thinking about the relationship between nature and the economy. The names we give to them highlight their characteristic features. All these concepts are normative, in that Neo-Aristotelian Scarcity and Utopian Scarcity were designed to serve as ideals to which people and societies should aspire, and Enclosure Scarcity was designed as a critique of the new commercial order. Throughout this chapter, as well as the rest of this book, we discuss different concepts of scarcity as parts of contemporary worldviews. These were the worldviews that explained and made sense of people's experiences and therefore shaped the way they went about their lives.

Neo-Aristotelian Scarcity

The dominant view among early modern European philosophers was that the relationship between nature and economy was best characterized as a delicate balance between environmental constraints and highly regulated human wants. Since there was little that could be done about nature's scantiness, it was essential that human desires for material goods were kept within carefully regulated limits. While merchants in Renaissance Italy and Antwerp made considerable fortunes, the dominant ideology of their respective societies was not founded on the idea of continuous economic growth. Instead, gain and greed were regarded as highly suspect. Firm restrictions on human passions were regarded as essential not only to a proper balance between economy and nature, but also to the stability of the social hierarchy, the backbone of the body politic. Far from the modern conceptualization of scarcity centered on unlimited wants and the endless pursuit of economic growth, the Neo-Aristotelian notion of scarcity focused on a balance of limitations.

While Enlightenment social theorists would later conceptualize society as the outcome of individual interests, rights, and actions, and would view society first and foremost as a vehicle for the pursuit of individual happiness, the fifteenth- and sixteenth-century European understanding of society was based on an entirely different set of principles. The focus was not on the individual (except in the spiritual realm) but on the well-being of the entire body politic. Effectively, individuals were invisible, other than in their capacity to contribute to the health and harmony of the broader community, whether that meant manor, town, or kingdom. While there were, of course, significant regional and national differences, we here seek to capture a set of general characteristics.

People and nature were commonly understood in early modern Europe as part of a Great Chain of Being, ranging from God to plants. Sir Edmund Dudley (1462–1510), for instance, regarded as the first Tudor political theorist, described how "god hath sett an order by grace bytwene hym self and Angell, and betwene Angell and Angell; and by reason betwene the Angell and man, and betwene man and man, man and beest; . . . which order, from the highest pointe to the lowest, god willyth us fyrmely to kepe withowt any enterprise to the contrary."[5] He called the human part of this hierarchy the Tree of Commonwealth. For this tree to grow strong and healthy it was essential that its five roots—love of God, justice, truth, concord, and peace—be firmly planted in the soil. It was not a coincidence that he used a metaphor that pointed to the close ties between the political, social, and natural worlds. The Prince was responsible, along with the clergy, for making sure that these values were properly nourished throughout society, so that harmony prevailed within and between social

The Great Chain of Being, 1577. The medieval and early-modern ideal of an ordered, hierarchical, and finite world was often represented as a Great Chain of Being, ranging from the realm of the divine to the lowest animals and plants. *Credit:* Diego Valades, *Rhetorica Christiana* (1579).

groups. The Prince was also responsible for nominating royal magistrates who conscientiously administered a just legal environment. The magistrate existed to "preserve and invigorate the hierarchy, to see that all ranks were filled, that justice was accorded to each member, that everyone performed his duties, and that no one was without the benefits appropriate to his status."[6] Next, it was necessary to have a powerful yet fair and generous nobility to manage the safety and administration of the commonwealth, and to serve as the primary landlord ruling over the peasantry. While they owned the land and received rent, landowners also had the paternalistic duty to alleviate the conditions of their subjects when times were difficult and food was in short supply. Dudley insisted that the nobility should operate as "the helpers and relevers of poore tenantes, and also be the manteynors and supporters of all poore folks in godes causes and matters."[7] Generosity and discipline served as the sinews of the social hierarchy. Finally, for the Tree of Commonwealth to remain vigorous there had to be plenty of industrious peasants, craftsmen, and merchants. The food, clothes, and instruments they produced and brought to market constituted the very lifeblood of the tree.

Society was thus seen as an integrated whole, with each segment of the population playing a specific and integral part in promoting the health and stability of the social organism. Sweden's first Lutheran archbishop, Laurentius Petri (1499–1573), explained how "each and every household should be governed in a Christian and proper way, how each and every one who belongs to the household, master, wife, children, and servants, should keep to their stations and engage with the other members in a manner that is compatible with God's commandments and the well-being of the household."[8] At a moment when memories of peasant rebellions such as the German Peasant Wars (1524–1525), Dacke's Revolt in Sweden (1542–1543), and Kett's Rebellion in England (1549) were fresh in people's minds, any movement for increased democracy, whether in the family, the local community, or in the broader polity, was considered seditious.

Since people had different roles and responsibilities, they had to be supplied with the appropriate resources to carry out their assigned tasks. For example, a peasant required a certain amount of wealth to maintain himself and his family to continue working the land and generating the requisite produce. A member of the nobility, by contrast, required a much greater portion of society's wealth because of his martial, ceremonial, and moral duties. It also cost him much more to educate his children so that they could shoulder the responsibilities incumbent on the nobility. The king and his court, needless to say, were entitled to the greatest share of society's wealth, so that they could uphold proper decorum, magnificence, and power, both spiritual and worldly.

With wealth distribution dictated by this ideal hierarchy, society was highly inegalitarian. Yet, because wealth was allocated according to function, a sense of justice prevailed. Instead of everyone receiving the same portion of society's wealth (arithmetic equality), people were allotted a share proportional to their station (geometric mean equality). As one historian points out, "only through geometric proportionality can equality be established between unequals."[9] As such, inequality was seen as constitutive of the social order and productive of the greatest possible common weal. Having a legitimate claim on society's wealth provided the poor with a sense of basic security, yet the size of the share was always at the mercy of the higher-ups: "All receive something, some more than others, according to the value placed on their contribution by those who receive most."[10]

What people could legitimately desire was thus determined by their places in the social order. No one, not even the king, enjoyed the privilege to completely unleash their desires, at least not according to the ideals prescribed by early modern philosophers. Desires were not individualized, nor were they free to follow the imagination, which meant that the Neo-Aristotelian concept of scarcity was characterized by a relationship between desires limited by social norms and a highly circumscribed nature.

To aid in the management of people's desires, numerous instructional treatises were published. The influential Swedish statesman Per Brahe (1520–1590), for example, devised a checklist of priorities for heads of families. He suggested that the first concern of every landowner should be to set aside enough wealth to meet his obligations to the nation's military. Next, the master must budget enough resources to satisfy the household's yearly requirements for food and nourishment, clothing, and shelter. The master also had to ensure that enough materials were available to furnish fine garments for ceremonial occasions, proper for each person's age, rank, and station, and to appropriately host guests and visitors. Next on the list of priorities was to earmark a portion of the household's wealth to pay for the baptism, education, and marriage of children. The penultimate category included charity, good works, and entertainment. Last on the list was a surplus to be used only in case of accidents and hardship, such as bad harvests, fire, war, illness, old age, or imprisonment.[11] The challenge was to stretch existing resources so that as many needs as possible could be met. Household management designed to satisfy a limited set of needs, not insatiable consumerism or infinite accumulation, was the key concept governing economic life and thus the prevailing notion of scarcity.

A Life of Limitations

Let's now take a closer look at how one community managed the relationship between the economy and the environment. In the historian Eamon Duffy's account of Morebath, a small Tudor village in Devonshire, we are offered a glimpse of sixteenth-century rural life.[12] While there were, of course, geographical and regional differences in how rural people lived in Europe (which constituted approximately 85 to 95 percent of the total population at the time), Duffy's study captures features that were common to most agrarian economies. Houses, for example, were constructed to mediate between humans and nature. They were built with thatch and cob, a mixture of clay, straw, and gravel. As long as the structure was properly roofed and limewashed, the surface was hard as stone and could survive for centuries. But if rain seeped in, the building would melt. The interiors of houses were sparsely lighted. Windows, if there were any at all, were small and far apart. Glass was far too expensive for the common peasant family to afford, so windows had to be boarded up at night to preserve the heat. Wax candles were available, but at a high price, so they were primarily used for special occasions (tallow candles were available and less expensive, but they were malodorous and smoky). Family dwellings often had two or three rooms, with the family living quarters placed on higher ground and the livestock, who lived under the same roof, placed below on a slope to facilitate drainage. A moderately well-to-do farmer family would have owned a plow ox, a bullock, three cows, two calves, a mare and colt, thirteen sheep, a sow, a goose and its gander, two hens, and one cock.[13] The animals' presence in the house provided much welcomed heat during the cold months, yet added significantly to the already fragrant atmosphere.

The modern divisions between people and animals, humanity and nature, young and old, work and leisure were still centuries away.

People spent the bulk of their time producing food and nearly everything that was consumed came from the immediate surroundings. Legumes, root vegetables, greens, and fruits were consumed wherever grown or within distances that could be reached by wagon or boat before the produce spoiled. When animals were slaughtered, hunted, or trapped, nothing was left to waste. Fresh meats and fish were salted, dried, or pickled to last, while the rest of the animal was used to make stews, soups, sausages, and lard. Spices, such as pepper, cloves, ginger, cinnamon, nutmeg, and sugar, were imported from faraway lands and were thus out of reach for most people. To obtain some disposable cash, some farmers participated in local commerce, selling dairy products, eggs, meat, wine, and vegetables to nearby towns. An even smaller number sold grains, salted fish, and dried meats to merchants who transported these outside the region, thus creating a rudimentary network of trade in basic foodstuffs. Most people and most foods, however, remained close to their places of origin.

Clothing was also derived from the immediate environment. The sheep's woolen fleece was sheared, cleaned, carded, spun, woven, and finished to produce a coarse wool that could protect the wearer from the elements. A peasant's wardrobe would consist of only a few garments. The design was simple and functional and colors, if any, were dull or faded. Dyes generating vivid colors, such as indigo, cochineal, woad, and brazilwood came from overseas colonies and were thus, for most people, prohibitively expensive. While peasants' day-to-day garments were simple and rustic, an individual would generally own one set of clothes for more festive occasions. These were frequently handed down from one generation to another. As a result, sartorial standards remained largely static throughout the early modern period, at least among the common people.

Occasionally, festivities and celebrations interrupted the meager experience of rural life. Key moments of the religious and agricultural calendar were marked by lavish banquets. Echoing fantasies of abundance in the folkloric land of Cockayne, where the habit was to "drink and eat / Freely without care and sweat / The food is choice and clear the wine," farming people put on their most splendid garb and consumed their finest proteins, rarest spices, and best liquors while engaging in revelries.[14] These occasions often provided opportunities for the higher-ups to treat commoners. For example, the wedding feast for the daughter of Elizabeth I's chief secretary of state, Lord Burghley, included 1,000 gallons of wine, 6 veal calves, 26 deer, 15 pigs, 14 sheep, 16 lambs, 6 hares, 36 swans, 2 storks, 41 turkeys, more than 370 poultry, 109 pheasants, 277 partridges, 615 cocks, 485 snipes, 840 larks, 71 rabbits, 23 pigeons, and 2 sturgeon.[15] Although feasting was increasingly frowned upon after the Reformation, it remained sanctioned by the church as long as it was kept within limits.

While feasts captured the imagination and provided welcome relief, we should not assume they were the center of peasant culture. Everyday life had its own rewards. Indeed, in a world of sharp constraints, simple material things and daily rhythms acquired profound importance. The case of Morebath offers a good example of this phenomenon. Devon was famous for its sheep breeds. By the late sixteenth century, a thriving wool industry developed in the county. The profits from the wool trade found their way into the parish economy of Morebath. Of special significance were the candles burned on feast days in front of the images of the saints and the Virgin Mary in the parish church. To keep a store of wax candles might seem an almost trivial task for people living in modern abundance, but to the parishioners of Morebath the funding of the candles required considerable effort and organization. The candle store of Our Lady (the Virgin Mary) became a collective responsibility for the village. A flock of about forty sheep produced the funds to pay for these candles. The two wardens of the store of Our Lady distributed the sheep to different households around the village who had to care for the sheep over the course of the year. At any given point, about two thirds of the families in the parish kept a sheep from this flock. Every year, the vicar of the parish held a great reckoning when he made a public account in great detail of the whereabouts and well-being of Our Lady's sheep over the past twelve months. He noted which sheep had fared well and which had been neglected or died. The sheep count thus served not just as an audit for church income but a moral lesson to the parish about the contribution of the individual to the community and the spiritual meaning of livestock management.[16]

The story of Morebath's sheep shows how moral imperatives shaped the interaction between economy and nature. Far from feeling helpless in their struggles with nature, early modern men and women gained a certain sense of control and dignity from their religious faith and spiritual beliefs. A shrine or place of worship or sacrifice provided a locus of community and identity. Not all religious movements received the blessing of the Church. Alongside official religious worship, many regions maintained their own cultures of magical beliefs, from the herb lore of wise women to the laboratory of the alchemist. In one particular instance, uncovered by the historian Carlo Ginzburg, a group of "good-doers" or *Benandanti* in northeast Italy claimed the power of magical flight. At night, they traveled out of their bodies to do battle with witches using fennel stalks. The victory over the witches ensured good crops in the coming season. Such ecstatic experiences, Ginzburg argues, was part of a Europe-wide worship of fertility, echoing ancient beliefs in the earth as a mother.[17]

Yet spiritual beliefs were a double-edged sword. Early modern people had good reason to fear the natural world in its destructive aspect. They knew intimately the inhuman side of the environment: disease, famine, floods, and other disasters. Such events were never simply "natural" disasters but always seen as carriers of moral and supernatural meaning. The enchanted realm of magic and religious belief inspired hauntings, omens, and horrors. In the popular imagination, the benign visage of nature could swiftly turn into a nightmare landscape filled with darkness, monsters, and suffering. This was also the realm of divine vengeance, where an angry God stalked the sinners, avenging infractions with harsh blows. The hierarchical cosmos of Neo-Aristotelian Scarcity was fueled as much by fear as by hope. Moral corruption and social collapse went hand in hand with natural disaster and supernatural phenomena.

Nowhere in English literature is the punishing side of nature more apparent than in the tragedies of William Shakespeare. In *King Lear* (1606), the monarch falls into poverty and madness when his family and subjects betray him. In a pivotal scene, Cordelia discovers her father stark raving mad in a wheat field, carrying a crown of weeds made of darnel, nettles, and hemlock. For a society haunted by recent harvest failure, Lear's crown signaled the double breakdown of the monarchy and the natural order.[18] Three years prior to the publication of *King Lear,* in a letter to King James, Francis Bacon articulated a similar correspondence between the laws of nature and government: "There is a great affinity and consent between the rules of nature, and the rules of policy: the one being nothing else but an order in the government of the world and the other an order in the government of an estate."[19]

The Looming Threat of Commerce

Reflecting the social conditions in places like Morebath, Neo-Aristotelians viewed the balance between humanity and nature as the key ingredient of a healthy and stable society. In addition to putting strict limitations on human desires, it was necessary to properly distribute society's wealth throughout the body politic. This is where commerce entered. Just as the human body required good circulation among its vital organs to maintain a balance between the four humors (blood, choler, phlegm, and bile), without which, according to Galenic medicine, the body would become sick, there had to be a functional circulatory system in the body politic for resources to reach their proper destinations. This was the merchants' responsibility. A former merchant himself and one of England's first economic writers, Gerard Malynes (1585–1641), clearly recognized the importance of commerce to the well-being of society. In all "common-weales," he argued, there were six primary members, each with their own unique responsibility. It was the duty of "Clergie-men, Noble-men, Magistrates, Merchants, Artificers, & Husband-men" to instruct, fight, defend, enrich, furnish, and feed the commonwealth. By distributing commodities throughout society, merchants made sure that each seg-ment received what they needed to carry out their responsibilities and to uphold the geometric equality on which the social hierarchy relied. Ob-stacles to circulation would lead to imbalances that eventually would re-sult in sickness and decay in the body politic. But as long as markets were free and open, and the integrity of the monetary system was upheld, and honorable, honest, and skillful merchants conducted trade, commerce augmented the stability of the body politic.

Philosophers dating back to Aristotle had defended commerce, provided it was kept within certain limits. Aristotle argued that it was simply impossible for a society to uphold justice without exchange. In com-plex, hierarchical societies with an extensive division of labor, there had to be a way to mediate between different classes and professions. Most crucially, qualitatively different goods had to be made commensurable. As Aristotle noted in his *Nicomachean Ethics,* all goods "must be equal-ized; and hence everything that enters into an exchange must somehow be comparable."[20] Money was the only instrument that could perform the magic of making a shoe and a house commensurable. By establishing the rate at which a shoe exchanged for a house, money also shaped the relationship between the shoemaker and the builder. Money thus mediated between things, people, and classes—an absolutely essential ingredient of a just society.

If people had used money exclusively to mediate exchange, the vast philosophical literature on the morality of money in the Western canon would never have appeared. But, as Aristotle lamented, money was Janusfaced. Alongside its capacity to establish commensurability and facilitate exchange, money had an entirely different existence and identity as the protagonist in the art of wealth-getting, or what Aristotle referred to as *chremastistike*. Money's nefarious alter ego was the primary instrument in the process of boundless accumulation. Uses of money to make more money, either by lending on interest or by purchasing goods for the sole purpose of selling them at a profit, as in retail trade, were regarded by Aristotle as distinctly unnatural. Consonant with his broader thinking, Aristotle insisted that everything in the world has a purpose, a *telos*. A shoe's purpose is to be worn on the foot as protection and comfort. So, when it is instead used for the purposes of infinite moneymaking, it violates its own nature and becomes unnatural, producing "riches of the spurious kind."[21] The degenerate merchants who carried out these trades, driven by a desire for accumulation and pleasure, continued to incense philosophers long after Aristotle. Instead of enjoying virtue and friendship, the pleasure-seeking merchants lived in violation of the good life. As their "desires are unlimited, they also desire that the means of gratifying them should be without limit."[22] This clearly was not possible in a world in which nature's limits were understood as rigidly fixed. Uncontrollable desires therefore held the power to destabilize any society.

While philosophers had long been troubled by the merchants' boundless quest for profits and pleasure, the fear in early modern Europe was that this rapacious and hedonistic ethic would contaminate the rest of the social fabric and sanction the acquisition of property beyond rank and station. Unleashing people's appetites and passions would lead to the destruction of the crucial yet precarious balance between nature and the economy. For Neo-Aristotelian Scarcity to prevail, the lessons that Icarus and Midas learned the hard way about pride and greed could not be forgotten.

The Ideal of Moderation

The Augustinian monk from Wittenberg, Martin Luther (1483–1546), submitted a scathing critique of the emerging commercial order and its effects on Christianity in the year 1517. In rebelling against the Catholic Church, Luther eventually launched a new Protestant alternative. While he objected most famously to the sale of indulgences, his broader economic complaint, which he articulated in subsequent years, was that merchants had ushered in a new era of greed and selfishness that made it all but impossible to uphold piety and morality, justice and the social order, and a proper balance between the economy and nature. As the Church failed to contain commerce, merchants were free to create a world in their image.[23]

Luther was far from opposed to commerce. Much like Aristotle and Aquinas, he viewed trade as essential to society. "One cannot deny," he noted, "that buying and selling are necessary. . . . For in this manner even the patriarchs bought and sold cattle, wool, grain, butter, milk and other goods."[24] The problem was that merchants were now operating in a culture devoid of any sense of self-restraint, self-control, and benevolence. They no longer asked how much their own goods were worth or how much effort, trouble, or risk they had incurred in procuring them; their only concern was to sell their goods as dearly as possible. This meant that the most relevant consideration to the merchant was the "other man's need; not that he may relieve it, but that he may use it for his own profit."[25] An utter disregard for the means, needs, and circumstances of one's brothers and sisters was certainly not an attitude becoming of a true Christian. Instead, according to Luther, when selling their goods, merchants should ask themselves how much trouble, labor, and risk they should be compensated for and how great a profit was compatible with "honest living." Such a calculation would generate the just price—the price that kept society fair and stable. Any price charged above this level would constitute "avaricious gain" and thus be a violation of decency and virtue.[26]

Infused with a desire to make as much money as possible, merchants all too commonly manipulated the market to establish unfair advantages. Luther was deeply troubled by this behavior.[27] He listed a number of methods whereby merchants established power over their customers. One of the most frequent strategies merchants used was to charge a higher price when they sold their goods on credit, thus taking advantage of the fact that the purchaser did not have ready cash on hand. They also charged higher prices when they were the only seller of the good in the region. Such monopoly power was particularly effective in circumstances when demand was inelastic—when, in Luther's words, "people must have it."[28] Unscrupulous merchants were constantly trying to establish such power. One popular strategy was to buy up all the available supply of a particular good to corner the market and create an artificial shortage. By engrossing all available goods, a seller could avoid competition and push prices up. Alternatively, the merchants could sell their goods below cost, thus forcing others to do the same. Having driven the smaller merchants out of business, the remaining rich merchant could then raise prices with impunity. Other sordid business strategies that he found in violation of proper ethical norms included the practice of a local merchant offering to sell goods he did not possess to a foreign merchant at a price higher than the market price.[29] The local merchant then bought the goods at market price and delivered it to the foreign merchant at a profit. Another strategy was to contact merchants who were under pressure from their creditors and offer to purchase their inventories at low rates. If the merchant in trouble declined the transaction, an associate of the fraudulent merchant would offer an even lower price. The expectation was that, after his inventory had been undervalued a number of times, the merchant would become anxious and sell below the market price, at a profit to the fraudulent merchants.

Luther revealed an intimate familiarity with all the different ways that merchants defrauded each other and the public, all of which were grounded in those merchants' failure to uphold the most basic of Christian rules: do unto others as you would have them do unto you. Monarchs should have tried to put a stop to such sordid business dealings, but instead they were often more concerned with trying to garner favors from the rich merchants. As Luther said of kings and princes, "They hang thieves who have stolen a florin or half-florin and do business with those who rob the whole world and steal more than all the rest." Recalling the words of the stoic philosopher Cato the Elder, he concluded: "Only the petty and stupid thieves lie in prisons and in stocks, whilst public thieves walk around dressed in gold and silk."[30]

The unscrupulous commercial mentality, Luther complained, "has gone so far and has completely got the upper hand in all lands," and, as such, merchants were no longer performing their vital function of circulating society's wealth.[31] The Church, he argued, had become overly permissive and accommodating of the merchants' antisocial behavior.[32] The time had now come to restore proper morality in the marketplace. Most fundamentally, Christian conscience once again had to govern the practice of buying and selling; without it, just prices would never be consistently upheld. Moreover, to infuse a sense of conviviality into commercial relations, it was necessary to reintroduce a general ethic of moderation. It was crucial, through moral suasion or laws, to manage "people's urges, passions, or appetites" to chart out a middle course "between excess and deficiency."[33] The seventeenth-century English bishop Joseph Hall, for example, proclaimed that the "chief employment of moderation is in the matter of pleasure, which like an unruly and headstrong horse is ready to run away with the rider, if the strict curb of moderation doe not hold it in."[34] Instead of indulging their appetites, people should be taught from an early age to value thrift, industry, perseverance, and contentedness. Excess and deprivation were cast as social pathologies, and *mediocritas* as the ideal.[35] As long as moderation was upheld, the relationship between nature and the economy would remain properly calibrated. This ideal concept of Neo-Aristotelian Scarcity was designed to influence people's understanding of the economic and natural worlds.

When moral suasion proved inadequate to rein in the passions, religious and secular authorities turned to sumptuary laws—which were devised to govern everything from dress and food to furniture and coaches. In the centuries labeled by one historian the "age of sumptuary legislation," early modern princes, magistrates, and church officials relentlessly sought to police people's appearances—in particular, the way they dressed.[36] Since clothing constituted the most visible marker of class, firm, detailed rules were deemed essential to avoid blurring distinctions. To dress above one's rank, counterfeiting one's true standing, was symbolically to subvert the social hierarchy. For example, in Tudor

England, to limit sartorial transgressions and keep people in their places, the law declared:

> None shall wear in his apparel:
> Any silk of the color purple, cloth of gold tissued, nor fur of sables, but only the King, Queen, King's mother, children, brethren, and sisters, uncles and aunts; and except dukes, marquises, and earls, who may wear the same in doublets, jerkins, linings of cloaks, gowns, and hose; and those of the Garter, purple in mantles only.
> Cloth of gold, silver, tinseled satin, silk, or cloth mixed or embroidered with any gold or silver: except all degrees above viscounts, and ciscounts, barons, and other persons of like degree, in doublets, jerkins, linings of cloaks, gowns, and hose.
> Woolen cloth made out of the realm, but in caps only; velvet, crimson, or scarlet; furs, black genets, lucernes; embroidery or tailor's work having gold or silver or pearl therein: except dukes, marquises, earls, and their children, viscounts, barons, and knights being companions of the Garter, or any person being of the Privy Council.[37]

Not all sumptuary laws were equally effective. Some legislation were met by resistance. For example, in Milan, gold and silk workers, embroiders, and merchants aggrieved by a 1565 ban argued that restrictions on certain garments, accessories, and jewelry severely hurt their trades and therefore the city. Some producers ingeniously circumvented the legislation by either lowering the quality of their goods or producing imitations with different fabrics, while others sought out foreign markets for their goods. The city authorities did their best to uphold the regulations. In Florence, for example, policemen patrolled the streets and fined anyone breaking the rules. Legal records indicate that people from all walks of life were subject to fines, from the city's elites down to butchers and barbers. In other places, such as Bern, Switzerland, fire wardens, clerk, and messengers acted as sumptuary officials and received one-third of the fine as a reward for policing the codes.[38]

Enclosure Scarcity

One of the most astute social critics of the sixteenth century, the English statesman and Renaissance humanist Thomas More (1478–1535), was convinced that the Neo-Aristotelian concept of scarcity, based on a carefully calibrated balance between a limited nature and bounded desires, had been made irrelevant by recent developments—in particular, by the nascent commercialization of Europe. Due to the expansion of commerce, Neo-Aristotelian Scarcity had simply become incompatible with the social and economic conditions prevalent in Europe. In his famous book, *Utopia,* More examined how the cultural shift toward boundless greed and insatiable desires had given birth to a new form of scarcity—the condition of Enclosure Scarcity, as we call it—that he found both distasteful and inhumane.

In the first part of *Utopia,* More explored the failure of the *mediocritas* ideal. He argued that private property, markets, and money had forced people into a powerful vice, in which a part of the population, the elite, was engaged in a boundless quest for wealth to gratify false pleasures, while the other part, the common people, was compelled always to want more because they lived in constant fear of poverty. The movements to enclose the land were largely responsible for creating this novel condition. In the old system of tenure, the nobility held the ultimate claim to the land, but the tenant farmers had strong use rights and rents were infrequently renegotiated. This provided most of the rural population with a great deal of security and confidence that they and their offspring would be able to subsist on the land across many generations. By the beginning of the sixteenth century, however, landowners were no longer satisfied with the revenues provided by the tenants' rent payments and instead switched from agriculture to sheep pasture, hoping that wool would yield higher profits. In the seventeenth century, a second wave of enclosure sought to eliminate the traditional open-field strip system in favor of the creation of large, consolidated plots on which more efficient agricultural techniques could be introduced. This transformation of the British landscape continued into the nineteenth century, sparking fierce resistance and protest along the way (as we shall see in Chapter 2 and Chapter 4).[39]

While the rich previously felt an obligation to take care of the poor during periods of hardship, the enclosures freed them from such obligations and enabled them to concentrate exclusively on their own enrichment. In the early enclosures, figuratively speaking, it was the sheep that pushed the rural population off the land. More painted a vivid picture of how the sheep had "become so greedy and fierce that they devour men themselves."[40] The landowners' quest completely transformed the relationship between humanity and the natural world. Land ceased to be the existential and spiritual foundation of the community, and instead was turned into exclusionary and alienable pieces of property, existing solely for the purpose of accumulation. In addition to losing their homes, family roots, and ancestral belongings, the former tenants lost their main source of sustenance. What remained for them, More queried, "but to steal, and so be hanged . . . or to wander and beg? And yet if they go tramping, they are jailed as idle vagrants. They would be glad to work, but they can find no one who will hire them."[41]

Propertied men were now engaged in a boundless quest for accumulation, while the rest of the people found themselves in a constant state of dispossession. While rich and poor experienced scarcity differently, they were nevertheless part of the same social dynamic. Anticipating Karl Marx by three hundred years, More recognized that the poverty of the landless was useful to the interest of the wealthy, who now had access to a cheap source of labor and captive consumers who had no other choice but to buy at the market price.[42] Poverty amongst plenty—previously an unthinkable condition—had now become an inescapable feature of the new world characterized by Enclosure Scarcity.

Utopian Scarcity

To reverse this catastrophic development, More suggested, required nothing short of a revolution. A simple redistribution of property would not solve the problem, because wherever the institution of private property was in place accumulation would always run amok. "However abundant goods may be," More reasoned, "when every man tries to get as much as he can for his own exclusive use, a handful of men end up sharing the whole pile, and the rest are left in poverty."[43] His solution instead was to eliminate the possibility of accumulation altogether by getting rid of private property. To answer the naysayers, who argued that a society without private property was impossible, as "every man stops working," More offered an elaborate description of the island of Utopia, a society that not only did not have a system of private property rights, but also had dispensed with markets, money, and commerce.[44] This depiction of a post-commercial world prefigured the hopes of many nineteenth- and twentieth-century revolutionary radicals.

More's fictional island was two hundred miles wide and contained fifty-four cities that shared the same language, customs, institutions, and laws. Everyone was trained in agriculture as well as in a trade, often the same as their parents', such as woodworking, linen-making, masonry, needlework, or carpentry. Diligence in one's occupation was a must; laziness and inefficiency would quickly get one into trouble. Still, More was quick to point out, "no one has to exhaust himself with endless toil from early morning to late at night, as if he were a beast of burden."[45] Indeed, six hours of work per day would suffice to produce enough material wealth to satisfy the needs of the island's population.

Three primary factors ensured that needs were always met. First, in Utopia, everyone worked. He contrasted this with those European societies in which priests, landowners, nobility, "lusty beggars," and a significant proportion of women were not engaged in productive pursuits. Second, the people of Utopia worked more intelligently and efficiently than their European counterparts. Although their brains were not superior to those of the Europeans, they were more inclined to learn new methods, techniques, and technologies. This was necessary as their "soil is not very fertile, nor their climate of the best." To compensate for these challenges, they engaged in hard work, developed new technological solutions, and pursued a deep understanding of nature's secrets.[46] With modest material desires and a relative abundance of necessities, the Utopians were liberated, to some extent, from pressing constraints.

The third, and most essential, reason the Utopians lived in a condition of relative abundance was their drastically different view of consumption. Contrary to Europe, where "many vain, superfluous trades are bound to be carried on simply to satisfy luxury and licentiousness," Utopians did not prioritize the enjoyment of material goods. Fashion and luxury were simply not relevant modes of self-expression. They all wore the same comfortable and functional clothes, with small variations depending on gender and marital status. Fabricated desires had no place. The Utopians thus embraced the Roman philosopher Seneca's declaration that "Natural desires are limited; those which spring from false opinions have nowhere to stop, for falsity has no point of termination." The story of luxury, he explained, is that first "she began to hanker after things that were inessential, and then after things that were injurious, and finally she handed the mind over to the body and commanded it to be the out and out slave of the body's whim and pleasure." As to "the proper limit to a person's wealth," his guidance was simple: "First, having what is essential, and second, having what is enough."[47]

In the absence of a constant preoccupation with consumption, which necessitated endless toil, the utopians enjoyed an abundance of free time. They used this time to pursue true pleasures. The first category of pleasures arose from the practice of virtue and the contemplation of the good life. These were, unsurprisingly, regarded as the highest pleasures imaginable. Pleasures of the body, in turn, were divided into two classes. The first class included eating and drinking, the elimination of excesses in the body, sexual intercourse, and the relieving of an itch by scratching; the second class included the pleasures that come from the enjoyment of art and music. While the bodily pleasures associated with the former were definitely real and genuinely enjoyable, More ultimately found them transitory and insubstantial and he therefore ranked the aesthetic pleasures higher. Indeed, no other creature, More reported, "contemplates with delight the shape and loveliness of the universe, or enjoys odours, or distinguishes harmonious from dissonant sounds," to the degree that Utopians did.[48]

Commerce, markets, and money had no place in Utopia. All commodities were transported to one of four central distribution sites. Here the heads of every household, after carefully calculating their families's needs, could avail themselves of the requisite commodities, entirely free of charge. There was no reason for anyone ever to take more than what was necessary to satisfy the family's needs, for the simple reason that they knew there would always be enough for everyone. Thus, there was no secondary market, no fear of future poverty, and no possibility that anyone would use commodities out of vanity—which More defines as satisfied not by what a person has "but by what other people lack."[49] The fact that people simply had access to all they needed profoundly influenced the Utopian economy because it eliminated any sensible rationale for the endless pursuit of wealth.

There were no reasons, either, to accumulate jewels, gems, and precious metals. The accumulation of gold had revealed itself to have such divisive effects on society that the Utopians decided to take steps to eliminate it altogether. Although they possessed vast amounts of gold, which they had accumulated to pay for mercenaries in the event of war, they assigned no particular value to gold. To show that there was nothing special about the metal, they used it to make chamber pots and forced people who had been enslaved as punishment for heinous crimes to wear gold earrings and gold chains. For them, it was human folly to covet gold. No one should "take pleasure in the weak sparkle of a little gem or bright pebble when he has a star, or the sun itself, to look at."[50]

The commitment of the Utopians to eliminate private property was so complete that they did not even allow private property in their own houses. People's houses and gardens were exchanged every ten years. Yet, the fact that they did not own their real estate forever had no impact on how they cared for it. In fact, they cared so much for their gardens that they competed with each other over whose was most beautiful. This was the only realm in which the Utopians appear to have engaged in competitive behavior. It had everything to do with the creation and celebration of beauty, however, and nothing to do with establishing or representing hierarchy.

More's vision of Utopia constituted a radical alternative to the commercial world that was beginning to emerge in early modern Europe. More promoted a completely different relationship between nature, humanity, and the world of goods. His ideal society was based on universal participation in work, the elimination of classes, the communal ownership of land and its products, the abolition of money, the development of knowledge to increase the yield of the soil, and the complete redirection of human desires away from the consumption and display of riches and toward the enjoyment of conversation, learning, and beauty. As such, his anticommercialism went far beyond that of Aristotle's aversion to *chrematistike* and Luther's complaint about fraudulent merchants. For More, when "money is the measure of all things, it is hardly ever possible for a commonwealth to be governed justly or happily."[51] For Aristotle, by contrast, it was impossible to uphold justice in society without money, for the very reason that it served as a "measure of all things."[52] More prioritized the elimination of money as one of the most critical steps, suggesting that if "money disappeared, so would fear, anxiety, worry, toil and sleepless nights. Even poverty, which seems to need money more than anything else, would vanish if money were entirely done away with."[53] People would then focus their attention on pursuits that were convivial and collaborative, not exclusionary and competitive. Togetherness rather than rivalry, friendliness rather than strife, sharing rather than accumulation, would follow. Once again, More echoed the Stoic philosopher Seneca, who instructed his followers: "Turn instead to real wealth; learn to be content with little and call out loudly and boldly: we have water, we have barley: we may vie with Jupiter himself in happiness."[54]

While Christian moralists such as Luther prescribed moderation as a crucial ingredient of a functional commercial society, More offered a more revolutionary approach. More's utopian world was not one of great material abundance, yet people experienced it as such because the elements of life that mattered to people were never in short supply. Even though the world of goods was, objectively speaking, far from endless, there was a subjective perception of abundance. We might call it self-imposed scarcity, but that does not quite capture the idea that a culture of consumption could be so fundamentally different that it did not require voluntary austerity. As such, More's proposal was so radical that some people viewed it as a practical impossibility, pointing to the etymology of the term *utopia* as "no place." Others emphasized a different reading of *utopia,* namely *eutopia,* which meant a place of pleasure, happiness, and fulfillment that might or might not be attainable. Far more influential was More's critical concept of Enclosure Scarcity, a condition his contemporaries could see in reality all around them. Indeed, his depiction of a society in which the rich ceaselessly pursue false pleasures and the poor always want more because they have none shaped the conversation about scarcity for centuries.

Conclusion

For the social hierarchy to remain intact and the body politic and nature to remain in balance, the mercantile spirit of infinite accumulation and insatiable desires had to be checked. According to the proponents of Neo-

Aristotelian Scarcity, because nature was considered largely fixed, there was simply no option but to keep demand for material wealth within limits. Once commerce started to become more prevalent, Luther insisted that the Neo-Aristotelian notion of scarcity could be upheld if merchants remained dedicated to proper Christian ethics. If people committed themselves to lives of moderation, they could live comfortably and virtuously at the same time that they contributed to a sustainable harmony between nature and economy. More, in contrast, was not hopeful that desires once unleased could be curtailed. Instead, he argued that it was necessary to remove the institutions responsible for promoting the culture of unlimited profit-making and boundless desires. In their place, new institutions were required to foster a new zeitgeist.

This chapter has examined three ways of thinking about the relationship between nature and economy: scarcity as a carefully calibrated balance of limitations; scarcity generated by an incessant quest for accumulation and a never-ending fear of poverty; and scarcity as a form of sufficiency, made possible by a radical change in the culture of desires. In all three of these versions of scarcity, nature was not absolutely fixed, but it was severely limited. These ways of thinking about scarcity were all grounded in broader ideological visions and political convictions. In chapters to follow, as we travel through history, the reader will recognize echoes of these concepts of scarcity, revised and reshaped to fit specific historical moments and their prevailing imaginaries.

2

CORNUCOPIAN SCARCITY

In the midst of the Little Ice Age, a small group of thinkers launched an audacious program to master nature and set humanity on a path of infinite growth and improvement. With a growing confidence in the prospects of science and technology, intellectual elites were starting to believe in the possibility of reversing the Fall of Man and thus restoring humanity's control over nature. It had long been thought that, while both innocence and dominion had been lost when Adam and Eve ate from the tree of knowledge, innocence could be regained through religious devotion—but now it was even conceivable that, through the advancement of the arts and sciences, the loss of dominion could be reversed. Savants and entrepreneurs alike believed that, by conscientiously employing empirical and experimental methods in the study of nature and systematically applying the resulting knowledge to the production of material wealth, they gained the power to push the limits of nature, perhaps indefinitely. This sparked an unprecedented spirit of optimism.

While many thought the famine conditions associated with the Little Ice Age and the devastating warfare of the seventeenth century signaled the end of the world, a new generation of reform-minded thinkers saw in the chaos and breakdown of political and religious authority across Europe an opportunity for radical change.[1] Growth, improvement, and refinement became the new mantra. Previously it had been regarded as essential to restrain human desires for the upkeep of social stability, but the reconceptualization of nature as inexhaustible obviated that need. Rapidly shedding its moral taint, unlimited desire now became sanctioned as the driving force behind the infinite improvement of nature and society. This new worldview thus gave rise to the novel conceptualization of scarcity that we call, slightly oxymoronically, Cornucopian Scarcity. Infinite nature and infinite desires came to be seen as two entangled or intertwined infinities. The perception that the world of goods was constantly expanding, on an infinite scale, yet was at any given moment inadequate to satisfy people's insatiable desires, generated this particular version of scarcity.

Botero and Bacon on the Betterment of Society

There were early signs in late-Renaissance Italy of the looming shift in focus from order and harmony to growth and expansion. The political theorist Giovanni Botero (1554–1617), most famous for challenging the ideas in Niccolò Machiavelli's *The Prince,* published a tract in which he explored the possibility of sustained economic growth. Did it stem from "the fertility of the soil," he asked, "or the industry of its people"? On the basis of his observations of the commercial cities of Genoa, Venice, Florence, and Milan, he answered unequivocally, "nothing is more important for causing a state to grow, and for making it populous and well supplied with everything, than industry and a great number of crafts."[2] While nature provided the raw materials, it was "human ingenuity"—the "skill and cleverness" of artisans—that transformed natural materials into an "endless number of artificial objects" to satisfy human wants. Whether wool, silk, or iron, the vast majority of value creation occurred after the "primary substance" had been removed from the sheep, the caterpillar, or the mine. In the latter case, the revenues accruing to the iron mines were small, "but an infinite number of people make their living from working the iron and trading in it; they mine it, refine it and smelt it, they sell it wholesale and retail, they make it into engines of war and weapons for attack and defence, into endless kinds of tools for farming, building and every craft, and for the everyday needs and innumerable necessities of life, for which iron is no less needful than bread."[3] The key to expanding wealth, Botero concluded, was thus to develop the kind of "skill and craftsmanship" that facilitated the production of an "enormous variety of wonderful works." More than the most lucrative mines in the colonies of Potosí and Zacatecas, industry and skills constituted the true sources of wealth—a point that John Locke and David Hume would later embrace.

Botero captured a process of improvement that was underway throughout Europe. As London quadrupled in size between 1550 and 1600, the town turned second-largest city of Europe was brimming with immigrants from the countryside and from across Europe and the Middle East, engaging in pinmaking, glassmaking, medicine-making, toolmaking, bookmaking, instrument-making, ship-making, and so on. Practical-minded men and women were investigating matter and energy in order to develop newer, better, and cheaper products. While students at the universities in Oxford and Cambridge remained mired in theological debates, Londoners "were busy constructing ingenious mechanical devices, testing new medicines, and studying the secrets of nature."[4] The progress achieved impressed locals and visitors alike. One particularly keen observer noted that the mechanical arts "grow and improve every day as if they breathed some vital breeze."[5] He marveled at how, at first, the inventions "usually appear crude, clumsy almost, and ungainly, but later they acquire new powers and a kind of elegance." What struck him the most was the spirit of "human progress and empowerment" that energized these pursuits.[6] Yet, as impressed as he was by these advancements, he believed they only scratched the surface of what human knowledge was capable of achieving. The tinkering mechanics acted on a partial, impressionistic understanding of nature. For human progress to emerge from its current state of infancy, science had to break free from the yoke of ancient authorities and chart its own independent, utilitarian course.

This observer was none other than the English statesman and philosopher Francis Bacon, known to posterity as the "father of the scientific revolution." While the epithet is surely an exaggeration, there is no doubt that Bacon played a significant role in establishing a new understanding of science and society. He himself did not make scientific discoveries or uncover any laws of nature, yet his vision of scientific improvement was nothing short of revolutionary. Bacon wanted to change fundamentally the way nature was understood and to overhaul all existing knowledge claims by subjecting everything in nature to rigorous scientific investigation. Although it would be a herculean task to overcome the "deeply layered ignorance of nature," ultimately the effort would be worth it many times over, he insisted, as it would revolutionize the way people lived. Perceiving the profound difference between "the opinions and fictions of the mind" that had governed humanity in the past and the "true and practical philosophy" that would create a new future would be, he said, like "awakening from a deep sleep."[7]

Equipped with the proper method, scientists would soon discover humanity's "real powers" and launch humanity on a journey of discovery. The inductive method would unveil nature's invisible particles and forces, and reveal the causal mechanisms at work in nature. This would give scientists the power to "generate and superinduce on a given body a new nature or new natures."[8] While he allowed that the capacity to transform nature had to be "within the bounds of the *Possible*," he nevertheless believed that humanity would soon be able to create things that had never existed before—indeed, that could not have been imagined. The future would not be a recurring circularity, in which the social conditions of the present were reproduced. Instead, the future would be linear, with each successive generation living better than the last, thanks to the material benefits made possible by the discovery of nature's secrets.[9]

Having acquired the godlike power to transform and master nature, scientists would be more valuable to society than politicians, Bacon predicted. While politicians might enhance public welfare, the benefits they generated tended to be short-lived. He contrasted these to "the benefits of discoveries" made by scientists, which lasted "for virtually all time" and could be enjoyed by "the whole human race." Moreover, he noted that the implementation of new laws and policies often "entails violence and disturbance," but that scientific discoveries "make men happy, and bring benefit without hurt or sorrow to anyone."[10] Bacon's ranking of scientists over politicians was particularly noteworthy considering his own personal history of public service during the reign of James I, first as attorney general and then as lord chancellor.

For the new scientific method to "endow human life with new discoveries and resources," it was essential that there be collaboration among all branches of science.[11] There had to be a constant exchange, a back-and-forth, between particular sciences and the overall system of natural philosophy. Bacon offered in *The New Atlantis* (1626) a plan for the advancement of science which in many ways prefigured the modern research university. Solomon's House, as he named it after the wisest of kings, contained a vast number of natural and built laboratories for scientific experimentation. For example, it included caves, up to three miles deep, in which scientists experimented with coagulations and refrigeration, mimicked how new ores originated, and experimented with how different metals could be combined to form composite materials. It also had high towers, some of which were built on mountains so that they rose three miles above sea level. From there, scientists studied how different materials responded to the altitude, as well how various weather phenomena were formed. They also had at their disposal both saltwater and freshwater lakes, which were used to study the interaction between water and different materials, and how water interacts with the atmosphere. Mineral wells allowed them to examine how vitriol, sulphur, and nitrate affected the human body and how they might prevent aging.[12]

Solomon's House contained specimens of most of earth's plants, herbs, flowers, and trees, as well as birds, fish, reptiles, amphibians, crustaceans, and mammals. The natural history of plants was one of its major preoccupations. In its vast orchards and gardens people investigated how to increase the fertility and yield of the soil, and how to combine different plants to form new ones with desirable and useful characteristics. Animals were kept for experimental purposes so that scientists could test medicines and surgical techniques before they were used on humans. They also experimented upon animals to see if it were possible to grow them larger or smaller, to make them more or less fruitful, to alter their colors, shapes, or instincts. They explored how animals could be made more useful to humanity, as better sources of meat or heartier beasts of burden. As they instrumentalized both flora and fauna, the scientists of Solomon's House exhibited the very impulses—discipline, control, dehumanization, and exploitation of nature—that philosophers Max Horkheimer and Theodor Adorno would much later associate with what they called the "dialectic of enlightenment," a historical process they blamed for ultimately producing the genocidal tendencies of the twentieth century.[13]

Scientists in Solomon's House also experimented with practical applications. They had access to brewhouses, bakeries, wineries, and kitchens to develop new and improved foods with better taste, nutrition, and medicinal properties. The refinement of the mechanical arts was a priority. Scientists developed and improved the production of all kinds of useful commodities, such as papers, linen, silks, and dyes. Sophisticated furnaces allowed them to refine metals and minerals, and perfect the art of glassmaking. They made magnifying glasses and telescopes of exceptional quality, which enabled them to see things invisible to the human eye and make further scientific advancements. They also excelled in the development of engines and machines, paying particular attention to the improvement of weaponry. These pursuits included successful experiments to launch objects into flight and to develop "boats for going underwater."[14]

To express gratitude and foster respect for scientific accomplishments, Solomon's House erected statues and provided important inventors with generous rewards. Statues were adorned with iron, silver, or gold, depending on their subjects' degree of importance. Citizens of the New Atlantis, young as well as old, sang hymns daily in praise of "God for his marvelous works . . . his aid and blessing for the illumination of our labours, and the turning of them into good and holy uses."[15] For Bacon, Solomon's House depicted the ideal research environment, situated in a pious society that properly respected science and incorporated its findings in its governance. If implemented in England, it would serve as an incubator for the infinite improvement of nature and thus enable an entirely different relationship between nature and the economy—the relationship that had previously characterized the Neo-Aristotelian version of scarcity.

Unlocking Nature's Infinite Storehouse of Riches

Bacon's vision and method promoted the spread and institutionalization of science throughout Europe in the seventeenth century, paving the way for the formation of the Royal Society of London for Improving Natural Knowledge (better known as the Royal Society) in 1660 and the Académie des Sciences in Paris in 1666. Before these state-supported institutions were formed, however, the Baconian program was operationalized by various privately organized scientific communities, including the Bureau

Jan Amos Comenius, 1657. The spirit of universal reformation, captured by some of Comenius's many improvement projects, informed the new era of infinite advancement and endless growth. *Credit:* Title page of Comenius, *Opera Didactica Omnia* (1657). Designed by Crispijn de Passe, engraved by David Loggan. Mannheim: Thesaurus eruditionis.

d'Adresse, founded by Théophraste Renaudot in Paris, and the Hartlib Circle, a pan-European network of improvers coordinated from London by the Prussian émigré Samuel Hartlib (1600–1662). The Hartlib Circle fused science, spirituality, and educational reform in an effort to launch an ambitious "universal reformation" project. Their aim was to promote the advancement of learning and harness its power to generate a "perfect reformation"—one that would put an end to the devastating religious warfare that had engulfed Europe since Martin Luther nailed his ninety-five theses on the church door in Wittenberg.[16] The English poet John Milton praised Hartlib, seeing him as a "person sent hither by some good providence from a farre country to be the occasion and the incitement of great good to this Iland."[17]

Hartlib was part of a vast network of intelligencers and reformers— also including the Czech philosopher Jan Comenius, the Scottish theologian John Dury, the American alchemist George Starkey, the English agriculturalist Gabriel Plattes, the German diplomat Henry Oldenburg, the English gentleman-chemist Robert Boyle, and the English physician-economist William Petty, to mention just a few—who shared a deeply optimistic vision of the future. The Hartlibians found a match between Bacon's ideas, Paracelsian alchemy, and Christian Millenarianism. This, of course, should not come as a surprise, considering Bacon's own embrace of both alchemical and millenarian thinking. Far from solely encompassing the art of gold-making, alchemy constituted a well-established branch of natural philosophy in early modern Europe. The alchemical worldview had been developed over centuries, drawing on Neo-Platonist and hermetic thinkers, including Hermes Trismegistus, Pico Della Mirandola, Theophrastus Paracelsus, and Joan Baptista van Helmont. According to the strand of alchemy promoted by the Swiss medical experimentalist Paracelsus, everything in nature was alive, growing, and constantly, albeit slowly, striving toward perfection. Copper was aspiring to become gold, the acorn was in the process of becoming a tree, and a boy-child would eventually grow up to become a man. Alchemists believed that it was possible to hasten nature's natural maturations. By mimicking the conditions that prevail in nature, they sought to accelerate nature's inherent processes. The alchemists moreover believed that, within certain limits, they could transmute matter, turning one material form into another. To learn about the properties and virtues of natural bodies, they conducted studies at the level of atoms. Once they had developed a map of natural bodies, they experimented by combining different substances and components, under different temperatures, pressures, and humidity, to see how they reacted. As the properties and virtues of different natural bodies were more or less attracted to those of other natural bodies, the alchemists mixed and matched to develop new and better materials. They carried out experiments to develop, for example, fertilizers that could increase agricultural yields, grains that were resistant to crop failure, pumps for the draining of mines, methods to lower the carbon content of iron to make it less brittle, and medicines for all sorts of ailments. Indeed, any knowledge that might generate more effective medicines or improve yields from plants, trees, flowers, grass, meadows, forests, and wastelands was worth pursuing.[18]

The alchemical worldview conceived of everything in nature—humans, animals, plants, and minerals—as part of an interconnected universe, which also included the celestial sphere's planets and stars. There was thought to be a special correspondence among the sun, gold, sulphur, and man, for example, and among the moon, silver, mercury, and woman. These were mystical connections between the material and the spiritual dimensions of the cosmos. Many understood nature to be suffused with divine spirits. To be able to comprehend and transform nature, it was therefore necessary for alchemists to be spiritually aligned with the forces and energies they were trying to harness. While the reader might find this perspective on nature incongruent with our understanding of modern science, we should recall that the hermetic tradition not only informed Bacon and the Hartlib Circle, but also played an important role in the efforts by pioneering chemist Sir Robert Boyle and pathbreaking physicist Sir Isaac Newton to understand the unobservable forces governing nature.[19]

Millenarian Christianity further fueled the culture of optimism among the members of the Hartlib Circle. When Christ reappeared on the day of judgment, it was believed, he would launch a thousand-year empire, during which his chosen people would come to rule the world. All righteous believers would then enjoy the kingdom of heaven on earth for the next ten centuries. Progress would not be instantaneous, though, but would be realized gradually through the advancement and dissemination of knowledge. The more people who contributed to and who could understand the language of science, the faster progress would unfold. Knowledge was a gift from God and thus belonged to all of humanity in common. It should therefore never be monopolized by anyone. Moreover, since nature was God's greatest creation, the search for scientific truth was inseparable from the pursuit of spiritual clarity. Just as various Puritan denominations saw work as a form of prayer, the Hartlib Circle viewed the advancement in natural knowledge as part of both economic improvement and spiritual salvation. As a servant to religion, science brought people closer to God. Knowledge of the divine also inspired an increased reverence for learning in general, strengthening both the intellectual and moral fiber of society.

Armed with a new scientific method that promised to yield a dramatic improvement in instrumental knowledge about nature and a new set of techniques that allowed humanity to harness nature's hidden powers, people started to look at nature in an entirely different way. Humanity was now in command of "infinite Meanes of Reliefe *and* Comfort."[20] Our *"Native Countrey,"* Hartlib argued, "hath in its bowels an (even almost) infinite, and inexhaustible treasure; much of which hath long laine hid, and is but new begun to be discovered. It may seem a *large boast* or meer *Hyperbole* to say, we enjoy not, know not, use not, the one tenth part of that plenty or wealth & happinesse, that our Earth can, and (*Ingenuity* and *industry* well encouraged) will (by Gods blessing) yield."[21] The earlier understanding of nature as stingy and scant, as discussed in Chapter 1, was now replaced by the image of nature as a cornucopia. With nature no longer subject to strict limitations, the timeline of progress became indefinite, if not infinite.

For some seventeenth-century thinkers, the deployment of knowledge to enliven nature's dormant wealth could best be compared to finding the key to nature's storehouse of riches. Another analogy used by alchemists was to describe the refinement of nature as akin to procreation; they often characterized the creation of new substances as the result of the union between nature's feminine and masculine forces. This idea of nature being transformed as the result of its inherent sympathies and attractions was fundamentally different from the way that Bacon visualized change. For him, improvement involved the use of experiments to wrest or pry the secrets from the bosom of mother nature, so that she could be subdued and subjugated. The Baconian project thus asserted masculine domination over feminine nature.[22] In Bacon's own words, scientific knowledge gives the scientist "the power to conquer and subdue her, to shake her to her foundations."[23] Furthermore, by suggesting that the "secrets of nature" are best revealed not by allowing it to follow its natural course but by subjecting it to the "harassment" or "torture" of the mechanical arts, Bacon undeniably linked the improvement of nature to patriarchal violence.[24] Although drawing extensively on Bacon's thinking, the Hartlibians were partial to the alchemical worldview. Seeing nature as an organic whole, they held that the transformation of the world had to occur in harmony with both the spiritual and material dimension of the universe and they did not use gendered language of violence.

In the tradition of Bacon's *New Atlantis* and Johann Valentin Andreae's *Christianopolis* (1619), Hartlib Circle member Gabriel Plattes laid out the group's utopian vision of how science could be used to expand the material limits of nature in *A Description of the Famous Kingdome of Macaria* (1641). Cast as a travel account from a visitor to a faraway island, much like Thomas More's *Utopia,* the text has the traveler describing how people in Macaria "live in great plenty, prosperitie, health, peace, and happinesse, and have not halfe so much trouble as they have in these European Countreyes."[25] The reason for the affluence was their dedicated embrace of improvement. The king had empowered his five councils (husbandry, fishing, trade by land, trade by sea, and plantations) to make sure that all resources were optimally employed. Taxes were used to finance land improvements and the construction of excellent roads and bridges, which together helped turn the island into a "fruitfull Garden."[26] To make sure that all land was properly utilized, people were prevented from owning more than they could reasonably improve. Any landowners who refused to implement all the latest improvement techniques were fined on an escalating scale, until finally they were banished. Public funds were also used to finance "strongly fortified" plantations abroad.[27] As these plantations developed and began to yield profitable exotic commodities, they would more than pay for themselves. Infinite improvement thus meant both internal and external colonization.

If England embraced the same scientific zeitgeist as Macaria, it would soon be able to "maintain double the number of people, which it doth now, and in more plenty and prosperity than now they enjoy."[28] Plattes thought that England was already moving in the right direction. He had declared earlier in *A Discovery of Infinite Treasure* (1639) that all of the "new inventions and improvements" currently underway would promote "the future happiness and flourishing" of England, quite likely making "this country the paradise of the world." Everyone stood to benefit, but none more than the "working poore," who would "be imployed in these new improvements, in such manner that they may live twice as well as they doe now."[29] The future prospects were excellent, to say the least: "as God is infinite, and men are infinite by propagation, so the fruits of the Earth for their food, and cloathing are infinite, if men will consent to put their helping hands to this commendable Designe."[30] The consistent use of the adjectives *infinite* and *universal* is revealing of the scale of the Hartlib Circle's ambition to extend the limits imposed by nature on humanity.

When it came time to develop a model for the Royal Society, a number of Hartlibians collaborated with the Swedish reformer Bengt Skytte to develop a proposal for Sophopolis, a city of knowledge, complete with libraries, museums, laboratories, printing presses, workshops, hospitals, and a botanical garden. Sophopolis was designed as a cosmopolitan campus welcoming scientists and improvers from all over the world. To facilitate the reconvergence of the knowledge scattered around the world after God punished humanity for erecting the Tower of Babel, it was necessary to develop a universal language—a project that Skytte had been working on with Comenius for some time. Universal language was key to the development of universal knowledge, which in turn was essential to the Hartlibians' ultimate goal of universal reformation. Although Skytte's proposal was never fully implemented, the Royal Society incorporated many of the ideas of the Hartlib Circle, and many of its associates.

Abundance for Some, Poverty for Others

The Hartlib Circle developed an elaborate plan for the infinite improvement of nature, one of the two strands of Cornucopian Scarcity. We will soon turn to the other strand, the legitimization of boundless wants, but before we can explore how it metamorphosed into a benign force, we first have to take a brief look at how many seventeenth-century European thinkers viewed property, markets, and money—in short, what we now call the institutions of capitalism—and how these institutions contributed to the idea of infinite improvement.

The seventeenth-century Millenarian vision of radical reform often included the abolishment of private property and the turning of the social hierarchy upside-down. Most of the Hartlib Circle members, however, were not inclined toward such subversive politics. They believed that the existing institutional framework was well-suited to the continued advancement of science and its translation into material abundance and moral refinement. Much like Thomas Hobbes, they endorsed the institution of private property as the foundation of modern society. Without such rights, Hobbes had argued, "there is no place for Industry; because the fruit thereof is uncertain; and consequently no Culture of the Earth; no Navigation, nor use of the commodities that may be imported by Sea; no commodious Building; no Instruments of moving, and removing such things as require much force; no Knowledge of the face of the Earth; no account of Time; no Arts; no Letters; no Society."[31] Improvement, broadly defined, depended on incentives for personal gain. Indeed, the seventeenth-century meaning of the term *to improve,* some historians suggest, was not just to make better, but to render more profitable.[32] To create sufficient incentives for scientific discoveries to be widely adopted and implemented, it was therefore necessary that the better part of the nation's arable land be turned into private property through enclosures. And in case the reconfiguration of nature into private property did not create strong enough incentives to persuade landowners to make the most of their land, Hartlib, like Plattes, believed it should be within the power of the state to force them to improve their land.

The Hartlibians also endorsed the use of money as an integral part of the improvement process. Unlike philosophers inspired by Aristotle's understanding of money, they did not express any serious ethical reservations about its use. The greater worry for them was that the shortage of gold and silver that had plagued northern Europe for at least a century constituted a barrier to further growth. They were concerned that there would not be enough money available to support the circulation of the anticipated abundance of wealth. Drawing on their alchemical expertise, the Hartlibians first tried *chrysopoeia,* or gold-making, to solve the scarcity of money problem. As their efforts to find the magic tincture repeatedly failed, however, they turned to another solution: paper money. As some of the first economic writers to propose a generally circulating credit-currency, they insisted that nonmetallic money could function just as well as precious metals. The key was that the paper money be backed by, or redeemable for, a concrete asset such as land or precious metal. Their suggestion was therefore to create banks that issued paper notes on the security of mortgages or fractional reserves of coin. Henry Robinson, a trusted collaborator of Hartlib, declared confidently that such a bank was "capable of multiplying the stock of the Nation, for as much as concernes trading *in Infinitum:* In breife, it is the *Elixir or Philosophers Stone.*"[33] Credit money thus served as an important feature of the new scarcity concept; to the extent that nature was infinitely expandable, money had to be, as well.

But far from everyone agreed that the institutions of private property and money were well-equipped to promote progress. Just as the sheep-pasture enclosures of the sixteenth century, undertaken to increase the production of wool, brought misery and pain to the rural population, the new round of enclosures, designed to foster agricultural improvement, were devastating to those evicted from or otherwise forced to leave their land. As people were divorced from the land their families had worked on, played on, prayed on, and survived on for generations, the growing surplus population was left with few options. They could try to find work as hired hands, although few landowners were hiring, or they could migrate to cities in hope of working as porters, dockworkers, and servants. Conditions in most cities, and particularly in London, were unsavory and insalubrious. Poverty and lack of sanitation led to consistently high rates of mortality—which skyrocketed every time the plague struck, as it did in 1603, 1625, 1636, and 1665. Many immigrants from the countryside spent only short periods of time in urban squalor before they moved on, perhaps trying their luck on the other side of the Atlantic. Yet, some of the evicted famers stayed on in the countryside, refusing to abide by the new property-rights regime and working hard to eke out a living by transforming unused wastelands into arable land.

Gerrard Winstanley (1609–1676), a member of the Digger movement—famous for their defiance of the enclosures and for their efforts to dig and sow St George's Hill in Surrey, just outside London—documented the resentment and revolutionary zeal of the dispossessed.[34] He argued that private property was first and foremost a legal construct devised to exclude the vast majority of the population from access to land and the sustenance it generated. Far from a pathway to general abundance, private property would bring only inequality and greater poverty. God had not given the earth to a small minority of people, so that the rest of the population could become "Slaves, Servants, and Beggers."[35] Instead, Winstanley insisted, God gifted the earth to humanity in common so that it could serve as a "Store-house of Livelihood to all Mankinde."[36] The only reason that people abandoned their access to land was because theft and murder forced them to do so. There were no social contracts or social conventions as far as the eye could see. This meant, in effect, that anyone with a deed to land held it "by the power of the Sword."[37] While he invoked the historical example of William the Conqueror's 1066 invasion to advance the argument, his accusation that property constituted theft was just as applicable to the enclosures then taking place throughout England, and to the appropriation of native lands in the Americas.

Money was also implicated in the violence at the heart of private property. Money, Winstanley argued, served as "the great god, that hedges in some and hedges out others," while commerce is "the great cheat, that robs and steals the Earth one from another."[38] Together, these institutions made "some Lords, others Beggers, Some Rulers, others to be ruled; and [made] great Murderers and Theeves to be imprisoners, and hangers of little ones, or of sincere-hearted men."[39] Private property and money should not be praised for bettering the lot of mankind, but should be cursed as a devilish and tyrannical system of oppression that imposed poverty and suffering on the masses. The Diggers therefore had every right, divine and human, to restore the land as a "common treasury to all," and to work it for the benefit of everyone, not just for covetous landowners.[40] It was the only way for society to avoid Enclosure Scarcity, to use our own term from Chapter 1—the kind of scarcity in which the rich constantly wanted more, because there was no end to their greed, and the poor also constantly wanted more, because they were always on the brink of starvation.

John Locke (1632–1704), one of the foremost seventeenth-century philosophers and a member of the English Board of Trade and Plantation, could not have disagreed more with Winstanley and his "ragged band" of Diggers. He believed that turning land into private property through enclosures was absolutely essential to making the most of nature and generating as much material affluence as possible. Locke employed the same biblical language as Winstanley at the outset of his deliberation on property, stating that God had given the earth to "Mankind in common."[41] But matters were not meant to stay that way, as God also gave humanity reason to make use of nature for their "Support and Comfort." Sounding very much like Bacon, Locke suggested that God, reason, and the inherent scantiness of nature required humanity to "subdue the Earth, *i.e.* improve it for the benefit of Life."[42] By mixing their labor with nature and transforming it into useful things, people acquired command—property rights—over the resulting commodities. Once possession was established, it was entirely up to the owner to decide how to use the property. A reasonable person would appropriate only enough of nature's bounty to satisfy his needs, as anything beyond that would simply spoil and such waste would be an insult to God's generosity. If everyone limited their appropriations to what they would actually use, there would always be plenty of land left over for others to use. For Winstanley, this would have been an ideal situation, but for Locke it fell short of the progress of which he believed humanity was capable. Having people produce only to satisfy their own needs would not unleash the full productive potential of humanity, and the limits of nature would not be extended.

Money was the solvent that changed everything, Locke insisted. When humanity agreed that "a little piece of yellow Metal" should be exchangeable for every commodity and have the capacity to store up value indefinitely, the limitation on property ownership was lifted.[43] There was now a rationale for people to produce endlessly, as the products could be sold in the market and the value stored in money. This new economic reality would result in a massive increase of production. One acre of enclosed, efficiently cultivated land, Locke estimated, produced ten times more than one acre of common land. As the owners of land produced beyond their needs, they added substantially to the total amount of wealth available to the community—a surplus Locke regarded as a gift to "Mankind."[44] To illustrate his point, he compared the conditions prevailing in modern commercial societies with those in America. While nature had furnished Native Americans liberally with "materials of plenty," the indigenous population did not enjoy "one hundredth part of the Conveniences we enjoy: And a King of a large and fruitful Territory there feeds, lodges, and is clad worse than a day Labourer in *England*."[45] Locke had little or no sympathy for the native populations of the Americas whose land was rapidly being appropriated by Europeans. Nor did he pay much attention to the inequality resulting from the enclosures in Britain, which More and Winstanley had found to be such an abomination. Displaced people, Locke assured, would be able to find work with sufficient wages to allow them to access food through markets offering it at very favorable prices. Any who did not find gainful employment should be forced onto his majesty's ships or into parish workhouses.[46]

Everything in nature was now, Locke argued, calculable and fungible. Although physically immobile, land had, with the introduction of money, become economically mobile. While the Hartlibians saw advancement in science as the most generative force, removing the distinction between the possible and the impossible, for Locke, incessant labor served as the main engine of the infinite improvement of nature.[47] The motivating force behind labor was the feeling of "uneasiness" shared by most people. For the poor, this was anxiety about "hunger, thirst, heat, cold, weariness." For the more affluent segments of the population, it was the pressure that came with their unyielding quest for "honour, power, or riches" and a "thousand other irregular desires."[48]

To be sure, Locke did not fully explore the social rationale for boundless accumulation, nor did he probe the underlying psychology of infinite improvement. For a deeper reflection on such questions we turn to one of Locke's interlocutors, Nicholas Barbon, and then to Bernard Mandeville (1670–1733), the philosopher who offered the most scandalous affirmation of unlimited wants.

The Infinity Zeitgeist

If the natural world was infinitely improvable, it followed that the need to limit desires was no longer pressing. Indeed, the idea surfaced in the second half of the seventeenth century that desires not only are irrepressibly boundless, but serve as an important force behind the improvement of nature. The seventeenth-century thinker who most clearly articulated scarcity in these terms was the English political economist and real estate magnate Nicholas Barbon (c. 1640–1698). Drawing on the works of philosophers such as Giordano Bruno and Henry More, who had recently theorized an infinite universe consisting of endlessly extended space and an unlimited number of atoms, Barbon embraced the concept of infinity in his political economy.[49] He saw limitless potential for humanity's improvements in the context of infinite nature, infinite desires, infinite money, and infinite time.

Barbon made his name and fortune at the crossroads of the new sciences, the expansion of commercial empire, and bustling London real estate scene. He studied medicine at the university in Leiden, where many prominent seventeenth-century intellectuals, including Descartes, Spinoza, and Mandeville, had trained. Once he returned to England, Barbon dedicated himself to the rebuilding of London, a city that had been devastated by two catastrophic events in as many years: the Great Plague of 1665 and the Great Fire of 1666. As an entrepreneur and speculator, Barbon actively participated in the modernization of London with an eye toward making the city the center of world trade. He took on great risk, made full use of the existing credit mechanisms, and played fast and loose with medieval zoning laws to put his urban renewal schemes into action. He experienced moments of great prosperity, flaunting his success with luxurious living and sartorial splendor, but also moments of indebtedness and insolvency. As part of his real estate endeavors, Barbon contributed to the formation of new financial institutions. He launched one of London's first building-insurance companies, insuring owners against fire losses, and opened two banks, the most ambitious of which, the National Land Bank of 1696, was designed to compete directly with the newly founded Bank of England (1694). In the last decade of his life, Barbon served as a member of Parliament and engaged in a number of political economic debates, in print and in person, most notably with John Locke on the issue of the 1696 Recoinage.

Barbon authored a number of tracts, the most substantial of which was *A Discourse on Trade* (1690), in which he laid out his vision for consumer-driven growth grounded in insatiable desires. He embraced the Hartlib Circle's perspective on nature as a potential cornucopia. He regarded nature's fecundity as "perpetual," by which he meant that it could continuously replenish itself: "Beasts of the earth, Fowls of the Air, and Fishes of the Sea, Naturally Increase. . . . And the Minerals of the Earth are Unexhaustable." Through art and ingenuity, people were able to transform nature's abundance into an ever-expanding world of goods, and "if the natural Stock be Infinite," he pointed out, echoing Gabriel Plattes, "the Artificial Stock that is made of the Natural, must be Infinite."[50] Although Barbon did not agree with Locke on the primacy of labor, they shared the view that nature would not expand infinitely without a system of property, commerce, and money. The money supply, he concurred with the Hartlib Circle, had to be elastic to accommodate the expansion in wealth. Land banks, of the type he himself had launched, were, in his opinion, the best option. When paper money was issued exclusively on the security of land, each paper note was backed by a solid asset and the quantity of money in circulation could expand in proportion to the value of the nation's landed property, which gradually increased as a result of continuous improvements and population growth. As long as confidence in the issuing bank remained high, its paper notes would circulate without a problem: "Credit is a Value raised by Opinion; it buys Goods as Mony does."[51]

The most central and innovative part of Barbon's argument, however, was his claim that the fundamental driving force behind the infinite improvement of nature was humanity's boundless desire for consumption. Barbon joined contemporaries, such as Johann Becher in Germany and François de la Rochefoucauld in France, in arguing a position frowned upon by generations of philosophers that human desires are not only impossible to master, but should be recognized as the lifeblood of commercial societies.[52] While competitive consumption had earlier been the exclusive purview of the nobility, exemplified most famously by the spectacular displays at Louis XIV's court at Versailles, the growing middling class was now increasingly participating in the process and thereby, according to Barbon, contributing to the rapid expansion of the economy.

Barbon's impressions of the economic conditions around him were accurate. Standards of living in large parts of Europe began to improve significantly by the second half of the seventeenth century. An increased availability of goods, combined with higher incomes, allowed a rising middle class to live in more spacious and well-appointed houses, to dress in new materials and designs, and to enjoy a much more varied and palatable diet. It is estimated that England's gross domestic product doubled in the seventeenth century and that average incomes rose by fifty percent in the second half of the century.[53] The availability of exotic colonial goods played an important role in creating a sense of excitement about consumption. Asia, Africa, and the Americas figured in the imagination of Europeans as treasure troves yielding an unending variety of fascinating consumer goods. Sugar and spices made a huge difference in how food could be prepared and served, sparking a much more diverse culinary culture, as reflected in a significant increase in cookbooks published.[54] Sugar, coffee, chocolate, tea, and tobacco, most of which were produced by enslaved or otherwise unfree labor on plantations across the oceans, also promoted new forms of urban sociability and served as stimulants that enabled more persistent applications of industry, fueling the so-called Industrious Revolution.[55]

Alcohol consumption was up drastically throughout Europe. An Englishman, for example, imbibed on average some four hundred pints of beer per year. Beer was regarded as a sound and healthy drink, and along with roast beef it became a symbol of England's prosperity, as witnessed by the famous Hogarth print *Beer Street*. Colonial imports also multiplied the possibilities for sartorial expression. New styles from all over the world, combined with easier access to materials such as cotton, silk, and beaver pelts and dyes such as cochineal, brazilwood, and indigo, made for more dazzling and fast-moving fashion. The comforts and splendor of people's homes were also rapidly improving during this period. Indoor spaces benefited greatly from the better lighting and air quality made possible by chimneys and glass windows. As dwellings increased in size, desires for privacy led to proliferations of rooms, particularly in houses of urban mercantile elites. Household furnishings and housewares saw a remarkable rise as, for example, wooden platters and spoons were replaced with pewter, and luxuries such as mirrors and clocks became more common in people's homes. The new material culture fostered a sophisticated consumer, driven by the "propulsive power of envy, emulation, love of luxury, vanity and vaulting ambition."[56] Globalization, colonization, and slavery ushered in a fundamentally different culture of consumption.

Barbon recognized that consumption operated on a number of different registers. Most fundamentally, he made a distinction between "wants of the body" and "wants of the mind."[57] The former refers to all things necessary to support life, including foods, clothes, and lodging, which were everywhere dictated by a set of basic human needs—a certain quantity of calories and liquid per day to maintain healthy organs and certain qualities of shelter and clothing to keep body temperature within a certain range. These needs evolved, Barbon insisted, but so slowly that they could in practice be considered finite and absolute. Matters were different, however, with "wants of the mind," which would always be infinite and relative. Barbon explained that any person having reached a certain level of refinement in a commercial society naturally yearned for "every thing that is rare, can gratife his Senses, adorn his Body, and promote the Ease, Pleasure, and Pomp of Life."[58] This consumer particularly desired goods that signified distinction, riches, and greatness. Because commodities such as clothes, equipages, and houses served as "Marks of Difference and Superiority betwixt Man and Man," people desired these goods in large part because they conferred power, status, and prestige.[59] Luxury consumption was thus as much about the semiotics of power as it was about the pursuit of pleasure. Since power is always relative, the process of emulation, and in turn accumulation, had no discernible end. It went on and on, at an ever-expanding scale. While the competition for power always remained inconclusive, the real benefit of this process, Barbon argued, was the material abundance it engendered.

The incessant quest for recognition and social standing infused economic life with a ceaseless energy. By channeling self-interest into the sphere of consumption, a passion otherwise potentially destructive of morals and sociability was thereby transformed into a benign force, yielding benefits across society.[60] Fashion played a particularly important role in keeping people focused on consumption. It was, Barbon argued, "a great Promoter of *Trade,* because it occasions the Expence of Cloaths, before the Old ones are worn out: It is the Spirit and Life of *Trade;* It makes a Circulation, and gives a Value by Turns, to all sorts of Commodities; keeps the great Body of *Trade* in Motion."[61] As someone who himself had an acute sense of what was in vogue—a friend described him "as fine and as richly dressed as a lord of the bedchamber on a birthday"—Barbon was intimately familiar with the power of fashion to fuel what Joseph Schumpeter would much later call *creative destruction.*[62] By not fully exhausting the usefulness of existing commodities before buying new ones, people wasted wealth at the same time that they created it. While he noted that the most extreme dedication to consumption—prodigality—might be detrimental to the individual, it was always beneficial to society; "For what is Infinite, can neither receive Addition by Parsimony, nor suffer Diminution, by Prodigality."[63]

The clothing and building trades were the primary sectors driving the expansion of the economy. These two sectors yielded some of the most visible consumer goods and were thus particularly important to the process of conspicuous consumption. These sectors also constituted two of the most refined areas of production, involving a great deal of specialization and technological sophistication. Moreover, the building trades sparked widespread industry and employment, while the end product—houses—also attracted more people to the cities. As the inhabitants of a city increased, so "increaseth the Emulation of the People; and Emulation increaseth Industry, and Industry Riches."[64] There could not, therefore, be anything that "conduceth more to the present Prosperity, or future greatness of a Nation, [than] to incourage the Builders of Houses."[65]

Barbon acknowledged that prosperity in commercial societies relied on inequality. Only the rich, which according to Barbon constituted one in a hundred, were able to engage in competitive consumption, while the poor were compelled to toil by fear of poverty.[66] Arguing that economic growth was based on "Industry in the Poor, and Liberality in the Rich," Barbon openly acknowledged that the material abundance in capitalism was reserved for the minority.[67] Yet the fact that a growing middle class could aspire to elevated levels of consumption meant that the underlying incentive structure was robust. While hunger sometimes subsided and the poor momentarily had less compunction to labor, emulation "provoaks a continued Industry, and will not . . . ever be satisfied."[68] As long as desires kept on multiplying, industry and ingenuity would ensure that there would always be additional new trades to satisfy "wants of the mind." In particular, "Trades that are imploy'd to express the Pomp of Life, are infinite."[69] The rich were thus engaged in a perpetual motion, enriching themselves, their city, and their nation. Barbon concluded that there were no reasons for society to restrain its consumption by exposing the population to moral instruction or sumptuary laws. To Barbon, infinite desire was the necessary partner of infinite progress.

Barbon's efforts to naturalize insatiable desires and conspicuous consumption paved the way for even stronger, more scandalous endorsements, none more infamous than Bernard Mandeville's *Fable of the Bees: or, Private Vices, Publick Benefits* (1714).[70] A Dutch Calvinist who had relocated to London in 1693 after earning his medical degree at Leiden,

Mandeville was a specialist on the treatment of hysteria and hypochondria—illnesses that were then linked to "the anxieties associated with the medical effects of consumption." Quite literally, the consuming classes were "consumed by fashion." They were engaged in a "pernicious life-style, which sacrificed health to the household deity of fashion."[71] Beyond his well-regarded medical treatise, Mandeville wrote a number of essays attacking the prevalent hypocrisy undergirding modern society, from slavery and charity schools to prostitution and religion. He challenged those who criticized flourishing commercial societies for promoting morally indefensible levels of consumption. To wish for "a flourishing Trade, and the decrease of Pride and Luxury is as great an Absurdity, as to pray for Rain and Dry Weather at the same time."[72] Thus, Mandeville did not stray far from Barbon but agreed that emulation, pride, and envy constituted the quintessential impetus behind the wheels of commerce. Yet Mandeville elaborated further on these themes, praising self-interest in a way that was deliberately designed to provoke moral authorities. He argued that, in societies where avarice, prodigality, vanity, fickleness, and lust were liberated from civil laws and religious dictates, people enjoyed "luxury and ease," excelled in "science and industry," benefited from "better government," and experienced unprecedented population growth. "Thus every Part was full of Vice," Mandeville observed, "Yet the whole Mass a Paradise."[73] Through the magic of unintended consequences, "The worst of all the Multitude, Did something for the Common Good." In one of his most memorable stanzas, Mandeville continued,

> The Root of Evil, Avarice,
> That damn'd ill-natur'd baneful Vice,
> Was Slave to Prodigality,
> That noble Sin; whilst Luxury
> Employ'd a Million of the Poor,
> And Odious Pride a Million more:
> Envy it self, and Vanity,
> Were Ministers of Industry;
> Their darling Folly, Fickleness,
> In Diet, Furniture and Dress,
> That strange ridic'lous Vice, was made
> The very Wheel that turn'd the Trade.[74]

Mandeville explicitly confronted preachers, philosophers, and moralists who condemned human desires as dangerous and destructive. He described in detail the rapid decline that a commercial nation would face if somehow people managed to embrace moderation. No one would seek to out-do each other by erecting spectacular houses or spending lavishly on fashionable clothing. "As Pride and Luxury decrease," merchants and companies would disappear, leaving manufacturing, arts, and crafts neglected and dead. Soon, all the progress and improvement that commerce had generated would be erased, leaving the nation poor and defenseless. Restraining desires was thus a certain recipe for disaster. Only fools, he concluded, *"strive / To make a Great an Honest Hive."*[75]

Mandeville, like Barbon, believed that prodigality in a society supports a thousand inventions that provide gainful employment for its working people—inventions "that Frugality would never think of."[76] Meanwhile, he found no indications that even the "greatest Excesses of Luxury . . . in Buildings, Furniture, Equipages, and Clothes" did anything to corrupt a person.[77] While he acknowledged that it was in some ways irrational for a man to buy more suits than he needed, the fact of "our boundless Pride" and deeply felt need to be esteemed by others, it made perfect sense to covet yet another flamboyantly tailored suit. The impulse every person felt to "wear Clothes above his Rank" was particularly strong in large cities, wrote Mandeville, echoing Barbon, because a person may "hourly meet with fifty Strangers to one Acquaintance, and consequently have the Pleasure of being esteem'd by a vast Majority, not as what they are, *but what they appear to be.*"[78] The ability to play with, manipulate, and counterfeit one's identity and rank in the social hierarchy meant that conspicuous consumption reached deep into the middling classes, if not further. In contrast to Barbon, who focused on the consumption of the rich and the labor of the poor, Mandeville insisted that consumption constituted the universal motivation and industry was the price everyone had to pay.

Mandeville's argument that pride was the driving force behind capitalism, and that no society could enjoy commercial prosperity without the vanity and avarice engendered by it, was far from popular with his contemporaries. In contrast to the members of the Hartlib Circle, who believed that the moral purity lost at the Fall of Man could be reversed through the restoration of universal knowledge, Mandeville accepted that man was irrevocably fallen. For him, the challenge was not to strive for any transcendental virtues, which he believed existed only as a figment of the imagination, but rather to create the best of all *possible* worlds suitable to an admittedly imperfect humanity.

Boundless Optimism

The intertwined infinities of Cornucopian Scarcity reflected a palpable sense of optimism in Britain by the start of the eighteenth century. Science and technology were rapidly developing ever more powerful ways to transform nature into commodities. Networks of trade were distributing manufactured goods around the globe and forging a sophisticated, cosmopolitan consumer culture among the affluent classes of Europe, America, Asia, and Africa. As Europeans tapped into the abundance of cheap natural resources and cheap labor—mostly enslaved or forced—in the parts of the world they were colonizing, the vast profits generated helped to fuel the urbanization of Europe. Together, merchants, planters, and manufacturers managed the production of unprecedented economic abundance. But it was not just rising economic affluence that contributed to the widespread optimism in Europe. The new culture of credit also played a critical role.

The Financial Revolution, which came into its own with the formation of the Bank of England in 1694, introduced new ways for people, especially in the middle and upper classes, to conceive of the future. The values of financial assets—stocks, bonds, annuities, options—were almost entirely determined by people's expectations of the future. East India Company and South Sea Company shares, for example, trading actively in Exchange Alley, a stone's throw away from the Royal Exchange in London, oscillated in value depending on people's predictions of these colonial ventures' profitability. When times were good, investors projected prosperity onto the future, which led to an appreciation of asset values and thus the enrichment of the owners of the financial instruments. Almost magically, financial markets thus transferred wealth from the future to the present. In addition to compressing time, financial markets also compressed space. A French investor in the Mississippi Company, for example, could buy a share on Rue Quincampoix in Paris, and realize a profit from the colonial trade without ever putting a foot on a merchant ship or participating in the violence at the center of the Atlantic slave economy.

The massive credit expansion enabled by the Financial Revolution allowed states, corporations, and individuals to spend and invest wealth that did not yet exist. It allowed wars to be won and commercial endeavors to be completed, the result of which was the materialization of the anticipated wealth. As long as expectations continued to be optimistic, credit stayed afloat and economic growth proceeded apace. As Hartlib and Barbon had predicted, credit truly had the power to accelerate the infinite improvement process.

But far from everyone sang the praises of the Financial Revolution.[79] In his immensely popular morality tale, *The Adventures of Telemachus, Son of Ulysses,* François Fénelon (1651–1715) harshly criticized financial and mercantile wealth, suggesting that it, in impoverishing the countryside, undermined traditional virtues.[80] The preoccupation with luxuries in cities made everyone "live above their rank and income, some from vanity and ostentation, and to display their wealth; others from a false shame, and to hide their poverty." The compulsion to appear rich was so strong that people would do anything, including going into debt or cheating. "This vice, which draws after it an infinite number of others, is extolled as a virtue, so that the contagion extends at last to the very dregs of the people."[81] The end result was overgrown cities, full of debauchery, surrounded by a desolate, impoverished countryside. A region so corrupted, Fénelon wrote, "resembles a monster with an enormous head, but the rest of the body, for want of nourishment, is meager and overextended, and bearing no proportion to the head." A wise leader would prevent this, he advised, knowing the "true strength and wealth of a kingdom consist in the number of the people, and the produce of the lands."[82]

Another common critique of financial wealth was that it was not real but fictitious. Because of its intangibility, the fear was that it could be manipulated by stock-jobbers or that it might enable projectors— entrepreneurs with dubious intentions—to fraudulently appropriate resources for chimerical ventures. As the eighteenth century entered its second decade, both stock-jobbers and projectors enjoyed greater legitimacy and were increasingly regarded as integral to the new, growth-based economy.[83] Nevertheless, concerns about the precarity of financial wealth prevailed. To address such trepidation, Daniel Defoe—famous novelist but also a voluminous writer on contemporary political and economic issues—introduced the figure of Lady Credit.[84] He carried on a running commentary in his newspaper, *The Review,* on the variable mood, fickleness, and volatility of modern finance.

Drawing on a set of common gender stereotypes, Defoe highlighted the affective, sentimental, and emotive qualities of financial wealth. Like a woman, he argued, when credit was in good spirits, its capacity to engender wealth was astonishing, but as soon as the mood changed, which might happen on a whim, credit collapsed. Subject to bouts of melancholia, Lady Credit became faint at the mere sight of even the slightest danger. She was also prone to hysteria, the quintessential female ailment of the age. Mandeville, in his famous treatise on hypochondriasis, had argued that this disorder originated in the female nervous system. In Defoe's allegory, it was the feminized psychological dispositions—passions, sentiments, and imagination—that constituted the core of the new consumer and credit economy.[85] Traditionalists, whose concept of the body politic still centered on an inflexible hierarchy in which everyone had their specific places and responsibilities, were intrigued by the productive potential of credit, yet feared its capacity to spark disorder.

Defoe certainly did not want to abolish credit, but he used the analogy of Lady Credit to advocate rational and strong—in other words, masculine—management of its feminine tendencies. By controlling its volatility, it was possible to harness credit's productive capacity and use it for the enrichment of humanity. Not unlike Francis Bacon, who saw science as the instrument whereby rational men could force mother nature to share her abundant productive potential, Defoe insisted that rational and prudent men could limit the fickleness and instability of Lady Credit, and thereby transform her into a dependable source of prosperity.

Although far from universal, the condition we describe as Cornucopian Scarcity gained traction toward the end of the 1710s. Members of the elite and the middling classes throughout much of Europe, and particularly in England and France, were gripped by an exuberant optimism. Inspired by the formation of the Bank of England and the South Sea Company in England, the Scottish gambler and financier John Law saw the opportunity to launch a new financial system in France that could ease the burden of its massive national debt—the result of Louis XIV's lavish spending on luxuries and military endeavors. The idea was to capitalize on the public's fascination with the allegedly inexhaustible riches available in the colonies. Law's most important vehicle was the aforementioned Mississippi Company, which controlled the trading rights to France's vast North American colony, extending from the Gulf of Mexico and the Appalachians to the Great Lakes and the Rocky Mountains. He used shares in the company to purchase the outstanding, heavily discounted government bonds. By taking these off the market, he greatly improved the government's financial position. People rushed to exchange their government bonds for the new stocks, which kept on appreciating in value. Law also added additional colonial trading companies to the scheme, including the slave-trading Company of Senegal and Company of Africa, to stoke excitement. His efforts paid off. Enthusiasm for the Mississippi Company only grew and the price of its shares continued to rise. Around the same time in England, the South Sea Company, whose primary revenue-producing activity was to sell slaves in South America, began to see its share prices rise rapidly. There was seemingly no limit to how much people were willing to invest in these companies. A speculative frenzy soon enveloped London and Paris. Prices of Mississippi Company shares went from 500 to 10,000 livres. Everyone made money.

The general exuberance about the capacity of both economy and nature to expand infinitely paved the way for a massive stock-market boom. But soon, both the bubble and the worldview would burst. Investors' fortunes would come crashing down and so would the prevailing understanding of the relationship between economy and nature. For the remainder of the eighteenth century, Enlightenment philosophers would labor to develop a new political economy, one that embraced progress but restrained the most outrageous beliefs in Cornucopian science and the most reckless endorsements of epicurean self-indulgence and Midas-like greed.

Conclusion

The intertwined infinities of Cornucopian Scarcity—infinitely improvable nature and infinitely expandable desires—constituted a sharp break with the past. While the Finitarian Neo-Aristotelian Scarcity had been based on the assumption that desires and nature were both bounded and absolute, Cornucopian Scarcity exploded the boundaries and introduced the notion that the nature-economy nexus had the capacity to expand end-lessly. Indeed, most of the members of the Hartlib Circle, as well as Barbon and Mandeville, felt quite confident that economic growth would address most other problems, be they of political, social, or moral character. In many ways, Cornucopian Scarcity prefigured the concept of Neoclassical Scarcity, only that it was even more optimistic about the rate at which humanity could transform nature into consumables. But the line between Cornucopian Scarcity and Neoclassical Scarcity was far from straight; the scarcity concept would take on many different shapes during the intervening two centuries. As we explore in the next chapter, Enlight-enment philosophers tried hard to retain features of Cornucopian Scarcity, while at the same time attempting to remove some of its more unrealistic expectations of economic growth, as well as the most objec-tionable assumptions about human hedonism and greed.

3

ENLIGHTENED SCARCITY

n 1723, Bernard Mandeville was nervously pacing the streets of London, awaiting the verdict of the Middlesex Grand Jury. He stood accused of publishing ideas harmful to good morals and with a "direct Tendency to the Subversion of all Religion and Civil-Government."[1] The city he was walking through had recently been rebuilt after the Great Fire of 1666. The most illustrious construction to emerge from the ashes, St Paul's Cathedral, was now nearly complete; only a few minor details remained. Many other monumental buildings were erected around it, new wharfs were built along the Thames, and people of all classes flooded to the capital. London was brimming with prosperity. Vessels were daily unloading exotic goods from England's sprawling colonial empire and novel wares were arriving from workshops around Europe. A new kind of affluence was visible everywhere. Wealthy merchants flaunted their riches, enjoying the admiring gaze of strangers. What made this moment unique was that luxury goods—drams of rum, pouches of tobacco, cups of coffee— were now within most Londoners' reach. Many were even able to enjoy a few items of clothing made of silk or cotton, perhaps with some vivid colors made possible by new, imported dyes. The visual and culinary culture of Europe would never be the same. The commercial exhilaration enticed many merchants to try their fortunes in London's financial markets. They speculated on future profits by buying and selling stocks, bonds, and options in Exchange Alley. Even Londoners who were less affluent could participate in the burgeoning world of high finance, as pubs and ale houses sold fractions of corporate stocks, appropriately named shares.

London's new consumer culture was infused with a novel ethic of self-interest. That Mandeville had recognized this, and revealed the uncomfortable truth that vanity, prodigality, and pride now fueled the city's unprecedented prosperity, was the reason he stood accused of crimes. The first edition of his *Fable of the Bees,* published in 1714 during a moment of prosperity and optimism, had received little attention from moral authorities. By the time the second edition appeared, the atmosphere had been radically changed by a watershed event in 1720: the burst of history's first stock market bubbles. The crashing share prices of the Mississippi Company in Paris and the South Sea Company in London sent shockwaves that reverberated throughout Europe. While historians still debate the extent to which these collapses triggered a significant downturn in the economy, there is little disagreement as to their impact on the period's economic culture and thinking.[2] Suddenly, the ideas of boundless consumption and infinite improvement began to be ridiculed in newsprint, philosophical essays, and ballads, as well as the period's newest literary genre, the novel.

The Cornucopian zeitgeist that had only recently taken hold was now scorned on all sides. Critics regarded the optimistic projections of alchemy and science as castle-building in the sky, dismissed the innovations introduced by the Financial Revolution as chimerical, and condemned the new culture of insatiable desires as a sure path to societal disintegration. In short, they rejected the whole notion of Cornucopian Scarcity as intertwined infinites. Yet, as Enlightenment philosophers and political economists set out to develop a new theoretical approach to scarcity, they found it difficult to jettison altogether the expectation of indefinite improvement. To create a separation between themselves and their predecessors, they made theoretical distinctions along three major lines. First, although they recognized that people were drawn to consumption, Enlightenment thinkers insisted that it was possible for people to temper their selfish desires over time. Second, while nature was capable of providing humanity with a great deal of wealth, the *philosophes* downplayed nature's capacity for the kind of spectacular economic growth envisioned by the previous generation. Third, while they believed that the economy could continue to grow for generations, they acknowledged the possibility that growth might not continue forever. Cornucopianism still provided the foundation for their thinking—only now, the riches flowing from the horn of plenty were no longer considered quite as extraordinary, and nor was "haughty *Chloe*," to use Mandeville's name for the female consumer, thought to be quite as desirous. And appearing on the distant horizon was the possibility of a stationary state, which both David Hume and Adam Smith, the protagonists of this chapter, thought might be a rather happy state of human affairs. The resulting configuration of scarcity, which we call Enlightened Scarcity, was thus based on insatiable desires checked by internal mental restraints and a steady but gradual expansion of material wealth, at least for the foreseeable future.

A Critique of Cornucopian Scarcity

Pundits, novelists, and philosophers put the blame for the 1720 financial meltdown on the shoulders of people such as John Law in France and John Blunt in England for their respective roles in the rise and fall of the Mississippi Company and the South Sea Company. A great deal of culpability was also assigned to the general "spirit of the age" for enabling such excesses and exuberance. As part of this critique, both prongs of Cornucopian Scarcity came under attack: the belief that insatiable desires served as the engine of infinite economic growth, and the conviction that science, aided by credit, had the power to unlock nature's hidden treasures and thus produce an abundance of material wealth. We discuss these two critiques in reverse order.

In light of the fact that the 1720 crisis was triggered by the stock market crashes, it is not surprising that most of the vitriol was aimed at the recent financial innovations. The financial scheme engineered by John Law in France was the most elaborate of its kind, combining the supposedly abundant riches of France's North American colonial possessions with a set of sophisticated financial mechanisms. "Law's System" was advertised to the investing public as a philosopher's stone of sorts, capable of empowering the French state, stimulating the French economy, and

enriching French consumers. After investors initially embraced his sales
pitch, however, share prices of the Mississippi Company plummeted in the
spring of 1720, and Law went from being the most powerful man in France
to the most reviled. He was now seen as the devil, Louisiana as hell on
earth, and the alchemy of finance as nothing but black magic.[3] From across
the channel, the already famous novelist and essayist Daniel Defoe ridiculed Law's scheme, leveling the charge that "you only screw'd up the ad-
venturous Humour of the People by starting every Day new Surprizes, new
Oceans for them to launch out into; so supporting one Chimera by another,
building Infinite upon Infinite, which it was evident must sink all at last
into infinite Confusion."[4] The mocking repetition of "infinite" is notable:
while previously used to spark the imagination of unlimited progress, the
term was now a target of ridicule.

The problems with credit, Defoe argued, was that it was poorly understood. Those who tried to describe the phenomenon of credit often
ended up lost in a labyrinth of metaphors. *"Like the Soul* in the Body," Defoe himself had written, "it acts all Substance, yet is it self Immaterial; it
gives Motion, yet it self cannot be said to Exist; it creates *Forms,* yet has
it self *no Form;* it is neither Quantity or Quality; it has not *Whereness,* or
Whenness, Scite, or *Habit."*[5] When financiers like John Law launched projects without fully grasping the power inherent in credit, they were playing with fire. They failed to recognize the Daedalian quality of credit that
Jonathan Swift, Defoe's rival for the title of most-celebrated Augustan
author, described in his long poem on the South Sea Company debacle
(c. 1720):

> On *Paper* Wings he takes his Flight,
> With *Wax* the *Father* bound them fast;
> The *Wax* is melted by the Height,
> And down the tow'ring Boy is cast.
> . . .
> His *Wings* are his *Paternal Rent,*
> He melts his *Wax* at ev'ry Flame;
> His Credit sunk, his Money spent,
> *In* Southern Seas *he leaves his Name.*[6]

Most critics of credit did not want to see credit abolished but reined
in—controlled and regulated. Defoe, for example, insisted that credit could

be almost as solid as gold if it were limited to private contracts between honest people who made credible promises and always punctually honored them. Credit had the capacity to thrive as long as people exhibited self-control and honesty. The problem was that honesty, credibility, punctuality, and self-control did not come easy.

A Critique of Science

Along with the critique of credit, the other sources of the cornucopian growth ideal, alchemy and science, also came under scrutiny for promising impossible riches. The brave new world of abundance envisioned by seventeenth-century natural philosophers was vigorously challenged by, among others, Swift, in his iconic novel *Gulliver's Travels* (1726). Swift satirized Bacon's Solomon's House and the Royal Society, founded in 1660, comparing them to the fictional "Academy of Projectors" that Gulliver encountered on the island of Lagado. He described a sprawling campus with some five hundred separate laboratories, in which projectors, "violently bent upon . . . improvement," experimented with ways to refine agriculture, trades, buildings, and manufacturers.[7] As Gulliver toured the campus, it soon dawned on him that nothing the projectors worked on actually had any practical application. For example, he observed one man who had for eight years been engaged in a project to extract sunbeams out of cucumbers, and in the adjacent room he came upon another man trying to "reduce human Excrement to its original Food."[8] As he continued his visit, he witnessed another projector trying to develop a method for building houses starting with the roof and ending with the foundation. A fourth man was feeding colorful flies to spiders, hoping that they would spin a colorful web—a material that could then substitute for expensive silks colored with foreign dyes. Poking fun at the Hartlib Circle's initiative to create a universal language to enable the restoration of knowledge, Gulliver described an absurd project aimed at "entirely abolishing all Words."[9]

The seventeenth-century enthusiasm about the potential of science to facilitate extraordinary improvements may eventually have been vindicated by the march of progress, from the steam engine to electrification to nuclear energy to computers, but from the perspective of eighteenth-century natural philosophers, the vision of progress promoted by the Hartlib Circle was blown out of proportion. They did not believe there was anything like a God-given source code that, once cracked, would enable humanity to unlock nature's infinite storehouse of riches, nor did they accept the idea that extraordinary breakthroughs would occur at a moment's notice. As practitioners of science moved away from hermetic and millenarian prognostications of an impending kingdom of heaven on earth, it was increasingly acknowledged that scientific progress would take time and require systematic work. Everything in nature had to be observed and properly investigated, with armies of scientists—professional, as well as amateurs—drawing up inventories, classifying and categorizing, conducting experiments, and formulating explanatory theories. All of these endeavors required a herculean effort and would bear fruit only gradually.

This shift in perspective on science was part of a broader transition toward a materialist and a mechanical worldview. The earlier search for universal knowledge had been based on the assumption that nature was infused with various spiritual forces. Some natural philosophers believed that inanimate, as well as animate, objects had souls, while others, unwilling to go that far, still allowed that nonliving entities might possess some kind of energy or force. Alchemists, for example, as discussed in Chapter 2, understood the cosmos as consisting of a series of unobservable correspondences between the microcosm and the macrocosm. Any adept seeking to operate on metals therefore had to worry not only about the heat of the kiln but also about the alignment of the stars. The new materialist and mechanical worldview, by contrast, understood the universe as composed of dead matter organized within a rational, machinelike order. Devoid of any inherent powers of their own, the inert particles, or atoms, relied on external force for movement or change. Everything in this world happened not because of sympathies, antipathies, occult powers, or spontaneity, but as a result of regular, predictable laws.[10] The creator of this grand machine, God, was referred to as the great engineer, architect, or watchmaker. The laws of this system were not immediately observable to the human eye, nor were its ultimate causes susceptible to human knowledge, but both were ascertainable through observation, experiment, and calculation. Equipped with these tools, chemists, botanists, physicists, and others set out to investigate all of nature's inert particles, with the ultimate aim of finding out how they all fit together as part of the natural order. Using the machine, not a living organism, as the conceptual framework for studying nature, made it easier to imagine replicating or substituting nature with mechanical devices.

In the middle of the eighteenth century, the mechanical worldview started to encounter some pushback. The impulse to ground explanations of nature solely in formal mathematics and reduce all of nature's operations to simple principles lost some of its momentum. David Hume, for example, challenged humanity's capacity to understand causal factors at work in nature. A variety of other philosophers, scientists, and novelists popularized the notion that moral sentiments were necessary to grasp nature's secrets. As philosophers questioned the capacity of instrumental reasoning to serve as the dominant source of knowledge, sensibility became part of scientific inquiry, creating an intimate correspondence between the natural and the moral sciences. Nature was envisioned as teeming with "active forces vitalizing matter, revolving around each other in a developmental dance."[11] Instead of a sharp contrast between human agency and nature's passivity, these Enlightenment vitalists saw a partnership between humanity and nature, based on a sense of shared creativity. By imitating and harnessing nature, they believed, people could reach further than if they strove for the domination of nature. We see manifestations of this approach in the work of many influential eighteenth-century natural philosophers, including Carl Linnaeus, Comte de Buffon, and Alexander von Humboldt. It also constituted a major inspiration for Romanticism, explored in Chapter 4.

For science to serve as a catalyst for wealth creation, society had to dedicate resources to its pursuit. Realizing that science yielded important medical, architectural, military, agricultural, and economic benefits, national and local governments took active roles in setting up infrastructure, financing scientific laboratories and gardens, sponsoring journeys of scientists abroad, promoting publications, and rewarding scientific innovations with honors and prizes. Many scientists also actively recruited members of the broader public to participate in the investigation of nature. Every spring, for example, Linnaeus organized botanical excursions on Saturday mornings, called *Herbationes Upsalienses*. These outings were famous throughout Sweden and Europe, attracting as many as three hundred students, villagers, and visitors at a time. Accompanied by horns, drums, and banners, the crowd followed its charismatic teacher across fens and fields, through beech and spruce forests, eagerly awaiting his instructions. Participants were encouraged to keep careful notes to learn the basics of his classificatory system. Linnaeus hoped that his excursions would inspire amateur scientists to contribute to his ambitious project of classifying the nation's flora and fauna in their entirety.

The more knowledge that was accumulated about nature, the more people were able to turn nature into useful wealth. Everything, Linnaeus insisted, "whereby humans are nourished, clothed, and adorned, supplied and . . . yes, all things that fall into the categories of clothes, luxury, wealth, amusements, and necessities, have their beginnings and origins in nature's kingdom."[12] Scientific and economic knowledge were, from this point, joined at the hip. Future prosperity was fundamentally based on commerce and industriousness—hard work in the fields, mines, and factories, as well as in the laboratories, mechanical workshops, and botanical gardens. Together, commerce and industriousness had the power to keep pushing the limits of nature and generating standard-of-living improvements for the bulk of the population. The idea of a more sensible and measured cornucopianism was taking root in the intellectual discourse of most European nations.[13]

A Critique of Insatiable Desires

Jonathan Swift (1667–1745) not only had much to say about the false promises of abundance by scientists; in *Gulliver's Travels,* he also delivered one of the most memorable critiques of the modern, insatiable consumer. Swift viewed the limitless appetite for bodily pleasure and the endless drive to accumulate as odious qualities of modern people. If five *Yahoos,* as he called them, were given food sufficient for fifty, they would not peacefully eat just what their stomachs required. Instead, each individual would be "impatient to *have all to itself.*"[14] Even more unappealing was their "undistinguishing Appetite to devour everything that came their way, whether Herbs, Roots, Berries, the corrupted Flesh of Animals." They also coveted food that could only be acquired by "Rapine or Stealth at a great distance" rather than the much healthier and tastier local alternatives.[15] The *Yahoos* were also "violently fond" of gold and silver.[16] They would dig with their claws for days to find nuggets in the ground and, if successful, hide them carefully and guard them with their lives. Gulliver was told by his host that on the fields where the shining stones were found, "the fiercest and most frequent Battles are fought."[17] As part of their insatiable desires, the *Yahoos* were overly libidinous and lacked any sense of restraint. Gulliver offered an account of a frightening moment when he, while swimming in a river, was sexually assaulted by a female *Yahoo*.[18] Having encountered a wide assortment of creatures during his wayward travels, Gulliver found the *Yahoo* to be the most "indocible, mischievous and malicious," as well as the "most filthy, noisome, and deformed Animal which Nature ever produced."[19]

The seemingly irrepressible pleasure-seeking common to commercial societies also preoccupied Daniel Defoe. Two of his most iconic title characters, Robinson Crusoe and Moll Flanders, served as examples of people who fell victim to their own boundless ambitions and insatiable appetites. Defying his father's advice to be satisfied with moderate success, Robinson engaged in risky endeavors to "pursue a rash and immoderate desire for rising faster than the nature of the thing admitted."[20] Lured by the storied profits available in the Atlantic slave trade, Robinson set sail on the voyage that would soon end in a storm and leave him stranded on a deserted island for twenty-eight years. Moll Flanders, born into poverty and criminality, spent her entire life pursuing wealth and higher standing. Seduction, marriage, prostitution, theft, deceit, counterfeiting, impersonation, and child abandonment were among the many strategies she pursued to satisfy her inextinguishable appetite for riches. It was only when Moll arrived in the American colonies, and realized her avarice was the source of her hardships, that she turned her life around. Committing herself to industry, honesty, and frugality, she not only lived happily for the first time in her life, but also acquired considerable riches.[21]

Similar to Moll, Robinson was only truly able to remove himself from the "wickedness of the world" once he freed himself from the false pursuit of happiness through pleasure and accumulation.[22] He noted that even the "most covetous griping miser in the world would have been cured of the vice of covetousness, if he had been in my case; for I possessed infinitely more than I knew what to do with."[23] Realizing that the gold and silver he had rescued from the shipwreck was useless to him on the island, he was free to discover what his true needs were, and how to satisfy them. He found that he was surrounded by bountiful nature and that, as long as he worked diligently, used his ingenuity to construct tools and equipment, properly cultivated his enclosed land, and carefully harvested and stored the produce, he would live in relative abundance. Committing himself to a life of systematic industry, he conducted empirical observations, kept careful inventories and journals, prepared a balance sheet of his state of affairs, patiently experimented with different materials, meticulously recorded days, weeks, months, and years, kept track of the weather and seasons, and constantly busied his mind with studying his surroundings. By investigating nature inductively and empirically, he amassed considerable data from which he was able to infer patterns in nature that he could then exploit. He learned numerous lessons from simple trial and error, even leaving one of his failed canoe constructions next to camp as "a *memorandum* to teach me to be wiser next time."[24] In short, he systematically used science to transform nature in ways that enabled him to satisfy his authentic needs and steadily, albeit slowly, amass a small fortune, including both a country house and a seaside house.

The Robinson who left the island after nearly three decades in many ways embodied the characteristics of an ideal member of the commercial middle classes. He was ingenious, industrious, and innovative. He was also well aware of the extent of his own needs, which he was able to meet with a mix of industry and frugality. His moderation did not entirely extinguish his appreciation of "things," or make him wholly uninterested in aesthetics and beauty. For example, after giving an account of the functionality of his cap, waistcoat, and breeches that he had made from animal skins, he noted that "I must not omit to acknowledge" that they were "wretchedly made."[25] He described in detail the style of his "high shapeless cap," his short jacket of goatskin, his fashionable open-kneed breeches, and his chic umbrella, his favorite accessory. Yet, he concluded with a palpable sadness, "I had so few to observe me, that it was of no manner of consequence; so I say no more to that part."[26]

Defoe's characters, Moll and Robinson, successfully managed to transform their moral psychologies. By transcending their Mandevillian selfishness, which had caused them so much suffering, they enjoyed not only greater material fortunes but also honest friendship, conviviality, and social virtues. A gradual increase in wealth and a sensible moderation of desires constituted the ideal relationship between nature and economy. Scarcity was thus regarded as a constrained cornucopianism and a measured epicureanism.

Softening people's self-love was a central aim for many eighteenth-century philosophers. They were interested in finding the ideal balance between self-interest, regarded as the engine of economic growth, and self-restraint, seen as the foundation of social virtues. The importance of reining in self-interest had been emphasized before—recall the ideas of Martin Luther and Thomas More in Chapter 1—but in the past, the focus had mostly been on restraints imposed externally, by religion or law. Eighteenth-century philosophers looked instead for a mechanism internal to the human mind, a way to change people's psychological disposition so that they willingly checked their own self-interest. If this were achieved, humanity's expansive yet calibrated desires could be brought into balance with nature's gradually expandable abundance.

Anthony Ashley-Cooper, the Earl of Shaftesbury, well known for his writings and association with John Locke, argued that all humans possessed a moral sense, as a faculty of the mind or a general psychological disposition.[27] This sense instilled a motive in people to look beyond their own narrow self-interest and be mindful of other's well-being. Aided by education and refinement, people's moral sense had the capacity, he argued, to promote changes in manners and customs, and thus to foster a culture of politeness—an etiquette of gentle and respectful public conduct. The eminent moral philosopher Francis Hutcheson further developed Shaftesbury's notion of a congenital moral sense.[28] He argued that, in the same way that humans have a built-in aesthetic sense that allowed them to distinguish between beauty and deformity, they also had a congenital capacity to discern right from wrong. While stronger in some than in others, this moral sense moderated people's selfish desires and provided their minds with motives for other-regarding actions.

The English bishop Joseph Butler acknowledged in one of his sermons that self-love was a powerful motive, but argued that humans were also endowed with a conscience and capacity for reflection, and "this faculty tends to restrain men from doing mischief to each other and leads them to do good."[29] The fact that people remained constantly mindful of their own reputations also kept them from solely pursing their own narrow self-interests. Conscience and reputation thus helped people check, but not eliminate, their selfish desires. This meant that people could still be driven by a desire to consume, but it would never be their sole motivation. They also cared about the well-being of others. Finally, some philosophers, including René Descartes and James Harrington, considered reason a viable counterweight to the selfish passions, to the extent it provided the individual with an ethical roadmap. Because reason followed a certain set of rules, they argued, it was epistemologically more robust than moral sense and therefore offered more legitimate instructions to the mind. Whether the corrective force was moral sense, conscience, reflection, or reason, the task was similar: not to drive out all self-interest, but to curtail it enough so that people were able to moderate their desires for pleasure and accumulation. These ideas would play an important role in both Hume's and Smith's efforts to chart a middle course between Neo-Aristotelian Scarcity and Cornucopian Scarcity.

David Hume's Enlightened Scarcity

David Hume (1711–1776), the most celebrated philosopher of the Scottish Enlightenment, was the first thinker to use the term *scarcity* as the starting point of his social, economic, and political analysis.[30] In his ambitious efforts to bring Newton's scientific method to bear on what Hume called the "science of man," he assumed two conditions: that nearly all people, in all historical ages, shared the quality of "selfishness" or at least "limited generosity" toward others; and that the realm of goods was always characterized by "their scarcity in comparison of the wants and desires of men." The tension following from these two assumptions shaped every human society.[31] To affirm his point, Hume considered two hypothetical scenarios in which scarcity was absent: when nature gave rise to "extreme abundance" and when the human mind exhibited "perfect moderation."[32] In the first case, because the "unlimited abundance" of nature would be sufficient to satisfy even the "most voracious appetites" and "luxurious imagination," there would always be enough food and shelter for everyone.[33] In the second case, because people's minds were filled with boundless generosity and they had as much "tenderness" for others as they had for themselves, there would be no competition for resources or goods, and humanity would "form only one family."[34] Unfortunately, Hume lamented, "the common situation of society is a medium amidst all these extremes," in which we are "naturally partial to ourselves" and "few enjoyments are given us from the open and liberal hand of nature." Thus, humanity has never escaped, nor will it ever escape, the condition of scarcity.[35]

While Hume did not believe that humanity could ever transcend the condition of scarcity, he argued that capitalism had the capacity to alleviate it. By simultaneously moderating people's selfish desires and expanding the availability of material wealth, the grip that scarcity held on the population could over time be loosened. Hume offered an elaborate argument, the contours of which we sketch below, for how capitalism could reduce the tension between desire for and availability of material wealth.

The first question Hume investigated was how to keep society intact in a world of selfishness and scarcity. While Hume was more sanguine about humanity than, for example, Jonathan Swift in his portrayal of the *Yahoos,* he did not believe that people were born with a ready-made capacity for sociability, nor did he think that their moral sense was strong enough to check their desires. Instead, he argued, the only way for inherently self-interested people to become more sociable was to shift their focus from the short-term to the long-term. In the long term, every individual was best served by being part of a society in which everyone lived in greater affluence—which for Hume meant a society in which people had the right to own property, trade goods in markets, and use money as a store of value. Once people realized the benefits that accrued from these three conventions, they would honor other people's property, uphold exchange agreements, and maintain the trust upon which money was based. Altruism would not be required for people to restrain their immediate self-interest to a sufficient degree. Taking a longer view, they would recognize it was more beneficial to them to act in ways that kept these conventions intact.

Hume's experience of living in three of Europe's most vibrant metropoles—Edinburgh, London, and Paris—made him a great champion of commercial society. Never offered a university position, he had to make his life outside the ivory towers, among what he called "men of action." Gregarious and curious by nature, Hume spent much of his time in the company of ambitious members of the middling sorts. Hume expressed his admiration for them, in particular for the merchants, whom he called "one of the most useful races of men."[36] It was their ingenuity, industriousness, and risk-taking that laid the foundation for what Hume regarded as a new era of progress.

The thriving consumer culture discussed in Chapter 2 intensified in the eighteenth century, forging a truly cosmopolitan economy. As demand soared for coffee, sugar, tobacco, and chocolate, a complex system of mass production, mass enslavement, mass transportation, and mass marketing emerged. The imperial nexus was epitomized by the popular image of an English merchant entering one of London's many coffee shops, presenting a silver coin from Zacatecas to be served sweetened coffee from the Caribbean in a Chinese porcelain cup, and enjoying it with some Chesapeake tobacco.

In addition to the consumption of new stimulants, the global economy was shaped by a new sartorial culture. Hume fully embraced this new middle-class aesthetic. As captured by the renowned portrait artist Allan Ramsay, the Scottish philosopher favored the *habit à la française,* wearing a coat over a waistcoat and breeches. This fashion also called for a pair of silk stockings, a jabot, a linen or cotton shirt with decorative cuffs, and a cravat. Hume's coat was made of brushed cotton, imported most likely from India, and its bright red color came from a dye derived from cochineal insects in Mexico. It was embellished with a metal brocade, intended to look like gold. His jabot and cuffs were decorated with linen lace made in France.

Cotton cloth from India and China was in high demand in European cities, with certain kinds of muslins and printed calicoes especially coveted. Asian silks were also in vogue, not only for clothing but for upholstery and wallpaper; so-called Chinoiserie adorned many of the new Georgian townhouses in London, Edinburgh, and elsewhere. The dyes used to give these textiles their rich colors—including brazilwood, fustic, turmeric, cochineal, woad, indigo, logwood, and sumac—were also imported. Domestic manufacturers, seeing that consumers could not get enough of the new styles, set out to imitate the desirable foreign goods. Aided by import-substitution policies, European manufacturers produced knockoff shirts, cravats, skirts, and blouses. But instead of simply copying the technology used in Asia, they employed their own technologies in ways that would eventually pave the way for the Industrial Revolution.[37]

As a jovial and social *bon vivant,* Hume took great delight in all kinds of luxuries—fine garments, sumptuous food, and spacious living quarters—earning him the nickname of *Le Bon David* in the Parisian salons. He saw nothing wrong with indulging responsibly in luxury consumption, unlike some of his contemporaries. The famous critic of consumerism John Brown, for example, argued in 1758 that giving too much rein to the imag-

Allan Ramsay, *David Hume,* 1766. Hume insisted that continuous economic growth would not only improve standards of living, but also promote cultural, intellectual, and moral refinement. *Credit:* National Galleries Scotland. Bequeathed by Mrs. Macdonald Hume to the National Gallery and transferred.

ination, which "admits of no Satiety," would make people vain and effeminate. This would spell disaster for England, Brown predicted, as it vied for global hegemony with France—a nation that the English never tired of slandering for being effeminate—in the ongoing conflict later named the Seven Years' War. Hume, however, remained unconvinced by the arguments Brown and others put forth. For him, as long as people did not bankrupt themselves, become entirely obsessed with consumption, or turn into lazy profligates, there was nothing destructive about their consumption. Moreover, because luxuries motivated people to engage in industry, commerce, and the advancements of the arts and sciences, they played a critical role in generating both economic growth and moral refinement.

While it is not difficult to see that the revolutions in commerce, science, and industry contributed to material prosperity, Hume was making a more subtle point: These revolutions, by causing people to become more virtuous and their tastes to become more refined, also contributed to the refinement of morals.[38] Industry contributed to moral improvement by giving people a sense of purpose, instilling discipline, and creating a regularity of conduct. Conscientiously pursuing a profession could be a form of "habit" formation, which Hume described as a "powerful means of reforming the mind, and implementing in it good dispositions and inclinations."[39] Hume further argued that when a person is engaged in "honest industry," his or her "mind acquires new vigour; enlarges its powers and faculties."[40] Industrious people are not only better at their jobs, they also acquire the capacity for more refined thinking, which included moral reasoning. Hume's views on the virtue-inducing qualities of diligent work are so similar to Defoe's that one might suspect he had Robinson Crusoe in mind as he wrote.

Commerce also contributed to the refinement of people's minds. To be a successful merchant, as Hume knew firsthand from his short stint as a merchant clerk in the commercial hub of Bristol in 1734, it was necessary to develop an array of worldly knowledge. Merchants needed to be well trained in writing, arithmetic, accounting, measuring technologies, and the assaying of gold and silver. They also needed to know the law, customs, and financial systems in all the places they traded. This, in turn, required language skills. Learned in many different fields, the merchant became the new *renaissance man*. Serving as the catalyst for globalization, the merchants were also responsible for connecting people from different nations, cultures, and religions. As citizens of the world they would "flock into cities; love to receive and communicate knowledge; to show their wit or their breeding; their taste in conversation or living, in clothes or furniture."[41]

The middling sorts also advanced the refinement of the arts and sciences. Without specifying the exact links, Hume argued that the "same age which produces great philosophers and politicians, renowned generals and poets, usually abounds with skillful weavers, and ship-carpenters."[42] Once the "minds of men" become energized, they "carry improvements into every art and science." This creates a culture in which "profound ignorance is totally banished, and men enjoy the privilege of rational creatures, to think as well as to act, to cultivate the pleasures of the mind as well as those of the body."[43]

The pursuit of the liberal arts was particularly instrumental to people's moral refinement, as it promoted a "delicacy of taste." Hume noted that, among the segments of the population, those "in the middle station" between the great and the poor made up "the most numerous Rank of Men, that can be suppos'd susceptible of philosophy."[44] They had enough leisure to engage with what he elsewhere called "the beauties, either of poetry, eloquence, music, or painting" that "draw off the mind from the hurry of business and interest; cherish reflection; dispose to tranquility; and produce an agreeable melancholy."[45] He believed strongly that "cultivating a relish in the liberal arts" strengthened people's judgment and thus put them on a path to virtue. It was not so much the specific content of any one philosophical, historical, or literary text that promoted moral refinement, but rather the practice of grappling with serious and complex ideas.[46]

Unlike philosophical predecessors who emphasized politeness, moral sense, and reason, or his best friend, Adam Smith, who focused on prudence and self-command, Hume argued that virtue was also shaped by one's social standing. Those situated in the middle of the social hierarchy (including lawyers, physicians, and others active in everyday commercial life) had more leisure than the poor but more motivation to be industrious and learn than the rich. Thus, they could exercise not just the humble virtues of the former or the generous virtues of the latter but "every moral Quality, which the human Soul is susceptible of."[47] As such, they avoided destructive excesses of consumption, knowing that "feverish, empty amusements of luxury and expense" could never compete with the "unbought satisfaction of conversation, society, study, even health and the common beauties of nature." Hence, as Hume observed elsewhere, "industry, knowledge, and humanity, are linked together by an indissoluble chain."[48] They could avoid destructive excesses of consumption, knowing that "feverish, empty amusements of luxury and expense" could never compete with the "unbought satisfaction of conversation, society, study, even health and the common beauties of nature."[49]

Capitalism, for Hume, thus had a tendency to ease the tension between economy and nature. By producing greater economic abundance while simultaneously enhancing people's tastes and morality, it promised to make the experience of scarcity less severe over time. It did not condemn a society to ever-escalating levels of production and consumption, but allowed for growth to tail off in the future as people embraced new priorities, shifting their focus to the pursuit of higher pleasures. The notion that capitalism might produce more refined citizens, with elevated preferences, became a common theme in the centuries that followed. Later thinkers, including John Stuart Mill, Alfred Marshall, and John Maynard Keynes, articulated similarly hopeful ideas about economic growth and refined sentiments progressing together. It is far from improbable that they were directly inspired by Hume.

Adam Smith's Enlightened Scarcity

In one of the first letters that survives from Adam Smith's pen, written in late October 1741 when Smith was eighteen years old, he asked his mother to send him "some stockings . . . the sooner the better."[50] Smith had just begun a six-year term as a Snell exhibition scholar at Oxford University with an annual income of forty pounds. Rather than buy stockings in town, he asked his mother to send them from Scotland. Stockings were both a "necessary and conveniency" in eighteenth-century parlance—essential to protecting oneself in a cold climate but also a source of comfort and status. In later life, when Smith took up the position of customs commissioner in Edinburgh, he favored white silk stockings to go with his linen coat, knee breeches, buckle shoes, and beaver hat. By that point, Smith's attire, while not quite as elegant as Hume's, was considered sartorially refined. But in 1741, he was surviving on scholarship money and whatever his mother could spare. The previous two winters had been severe; across Europe, persistently low temperatures had led to bad harvests, high grain prices, and increased mortality rates. Smith himself described 1740 as a year of "extraordinary scarcity." He probably worried that the winter of 1741 would be as harsh as the last one.[51]

Smith's letter to his mother anticipated many of his enduring interests in political economy. For Smith, the benefit of labor and exchange lay above all in the production of material goods that satisfied basic needs and comforts like wheat, wool, and timber. Food, clothing, and lodging were the "great wants of mankind." For this reason, the cycles of the natural world were never far from his mind. While he celebrated the productive potential of the division of labor, his view of industry was rooted

more in agriculture and skilled manual labor than in large-scale factory work or power sources like water and coal. It was not until Charles
Babbage published *On the Economy of Machinery and Manufactures*
(1832) that the factory system truly entered political economy. For
Smith, the origin of wealth still lay in agriculture. For him, the surplus
of the land fostered the growth of towns, new trades, and new manufactures. From humble origins like the knitting of stockings, capital began
to accumulate thanks to the interplay of agricultural improvement
with urban industry. While Smith left his readers with plenty of reasons to be optimistic about future economic growth, he never let them
forget that human economies were always nestled within the limits set
by the natural world.[52]

Smith differed from Hume in ascribing a larger role to nature in de-
termining a nation's wealth. Smith wrote: "The land constitutes by far the
greatest, the most important, and the most durable part of the wealth of
every extensive country."[53] Without fertile soil, a nation could not feed its
population of laborers. In English history, agricultural improvement had
gathered momentum for more than a century, but in Scotland a much more
recent change was playing out over Smith's own lifetime. In the middle decades of the eighteenth century, Scottish farmers achieved massive increases in agricultural productivity, expanding cereal yields by as much
as a factor of three. Smith was very much aware of these changes. Around
the same time that he began sketching a preliminary draft of *The Wealth
of Nations,* Smith left his university position to become the private tutor
of the future Duke of Buccleuch, a member of one of Scotland's wealthiest
landowning families. While this new line of work took Smith on his first
trip to the continent, where he met luminaries of the French Enlightenment, the position with Buccleuch also gave him an opportunity to experience firsthand the reorganization of a great estate in the Scottish
Lowlands.

Nature was a gift to humanity that kept on giving. While labor and
science augmented the productivity of the land, nature's myriad processes
never stopped working in ways that benefited humanity. Wealth, for Smith,
began with soil fertility: "In agriculture . . . nature labors along with man;
and though her labor costs no expense, its produce has its value, as well as
that of the most expensive workmen."[54] This "free" gift of nature was the
source of a continuous surplus that made agriculture fundamentally

different from manufacturing production: "It is the work of nature which remains after deducting . . . everything which can be regarded as the work of man. It is seldom less than a fourth, and frequently more than a third of the whole produce. No equal quantity of productive labor employed in manufactures can ever occasion so great a reproduction. In them nature does nothing; man does all."[55]

Much of the inspiration for Smith's view of nature came from the French school of political economy known as the *physiocrats* (from the French word for nature's government). This movement emerged in response to a century of frustration with the French state's perceived prioritization of manufacturing over agriculture, cities over countryside, and merchants and manufacturers over nobles and farmers. In *Telemachus,* the epic fictional account of France noted in Chapter 2, François Fénelon lodged the complaint that Colbertism, the set of policies put in place by the powerful French minister Jean Baptiste Colbert, was wreaking havoc on the social order and undermined the nation's strength. Fénelon claimed that nobles and farmers lived in a sorry state of impoverishment, while city-dwellers reveled in luxuries. As wealth flowed into the cities and people became enslaved to false passions, traditional virtues were eroded and manners corrupted. Credit exacerbated the problem by allowing status-hungry individuals to finance their luxury consumption, only to end up in debt and misery. While there were many French thinkers, Voltaire and Melon included, who drew inspiration from Barbon and Mandeville and defended the new culture of luxury, the negative sentiments expressed by Fénelon lingered and eventually provided the moral foundation for the physiocratic program, headlined by Francois Quesnay, Victor Riqueti de Mirabeau, and Anne Robert Jacques Turgot.

The physiocrats put forth a theory that proclaimed that nature was the only resource capable of creating new wealth. In the spirit of vitalism, they saw nature as a vast number of benign processes spawning soils, mosses, grasses, plants, flowers, trees, fruits, berries, vegetables, and all the animals of the oceans, the air, and above and below the ground. This was anything but a mechanical operation that could be understood as a rational system; it was an organic world beyond human comprehension.[56] Instead of working to subdue or control nature, the physiocrats said, humanity should cooperate and enable it. This partnership would in turn bring prosperity to human society.

In value terms, land—the physiocrats' shorthand for nature—was the original source of all wealth. It provided for the subsistence of the laborers, yielded remuneration to the landowners, and produced a surplus, the so-called *produit net.* This surplus could be spent on manufactured goods or reinvested into agriculture. The proportions distributed to these two categories greatly mattered.[57] Because manufacturing did not add any new value—it was in their parlance "sterile"—it was essential that enough of the surplus was put back into agriculture. According to Quesnay's famous *Tableau Economique,* at least half of the surplus ought to go to the land. If the landowners failed to reinvest at this level, Mirabeau warned, "the death and extinction of Society and the human species" would ensue.[58]

To facilitate the flow of capital into agriculture, the physiocrats promoted the liberalization of the grain trade and the implementation of a single tax. They believed that free markets would prop up the price on agricultural goods and therefore make it more attractive for investors to commit their money to land improvements. A single tax on the *produit net* would simplify the tax-gathering process and free landowners and farmers from the myriad taxes and fees they now faced. By shifting the flow of wealth from manufacturing and commerce to the cultivation of the land, overgrown cities would retreat, while rural areas would once again flourish. This would put a stop to the rampant vice ignited by the quest for luxuries and restore the natural virtues associated with rural life. The physiocrats thus proposed their own form of Enlightened Scarcity: If enough resources were allocated to agriculture, the nation would not only enjoy the fullest benefits of nature's inherent bounty, but also see a reduction in people's preoccupation with excess luxuries. The tension between wants and wealth would be eased.

Smith agreed with the physiocrats that nature was a crucial source of value for the commercial economy, but he rejected their critique of manufacturing as a sterile form of production. It is telling that he chose to open *The Wealth of Nations* with a glowing paean to the productive potential of specialization and machinery, leaving his analysis of the foundational role of agriculture to much later in the treatise. Only in the third part of the book did Smith explain that the key to growth was the interplay between agricultural improvement and town manufactures. A fertile soil under cultivation produced the initial surplus that encouraged a population of workmen to develop and settle in towns. Over time, such urban

growth and agricultural improvement became locked in a virtuous circle
of mutually advantageous trade. For Smith and his friends, this synergy
of town and country was a way of life, not just an abstract model. In the
learned Edinburgh clubs Smith frequented, he befriended both scientists
and agricultural improvers. Among Smith and Hume's friends were the
geologist James Hutton, the judge Lord Kames, the chemist Joseph Black,
and the physician William Cullen: all of them were involved in agricultural improvement schemes. No doubt Smith would have heard more
than his share of talk about dung and turnips.[59]

   If soil fertility provided one pole in Smith's conception of Enlightened Scarcity, human desire offered another central topic of investigation.
Much like Barbon and Hume, Smith saw consumer demand as endlessly
malleable. While the needs of the stomach were limited, "the desire of the
conveniencies and ornaments of building, dress, equipage, and household
furniture seems to have no limit or certain boundary."[60] Nowhere was
this stated more explicitly than in his parable about utility and desire in
The Theory of Moral Sentiments (1759) concerning a poor man's son who
succumbed to excessive ambition. Although Smith was hardly a literary
stylist, in this passage he presented a kind of miniature novel, in the spirit
of Defoe's *Moll Flanders* or *Robinson Crusoe*. Enamored with the affluent
lives of the highest social classes, the son imagined that riches would
bring him tranquility and happiness. To improve his position, he threw
himself into the task of social climbing, ingratiating himself with his superiors and patrons. In the process, however, the son ruined his bodily
constitution and mental state. Smith observed that what the social climber
desired most was actually available to him from the beginning. He had
sacrificed the "real tranquility that is at all times in his power" for an
"idea of a certain artificial and elegant repose" that was completely out of
reach. Wealth and power, Smith observed, were "enormous and operose
machines" that produced "trifling conveniencies."[61] They offered scant
protection from the genuine frailties and needs of human existence. "They
keep off the summer shower, not the winter storm, but leave him always
as much, and sometimes more exposed than before, to anxiety, to fear, and
to sorrow; to diseases, to danger, and to death."[62] From this, Smith drew a
radical lesson. He suggested that all classes could enjoy the same "ease of
body and peace of mind," regardless of rank. A beggar could find as much
security in life as the great landlord. The key problem was to master fear

and ambition. Smith here came close to the ancient Stoic ideal of *apatheia;* mental composure and freedom from passions ensured the best state of mind.[63]

For all his cynicism about the empty fulfillment of material riches, Smith also defended the desire to imitate the wealthy because he recognized that such emulation produced advantageous consequences for society as a whole. A total abnegation of desires and passions might please the ascetic, but if everyone embraced this kind of virtue, their conduct would reduce the overall wealth of the nation. Smith agreed with Hume that vanity and ambition goaded mankind toward advances in the arts and sciences and greater conquest of the natural world. Providence dictated that the fortune of the few directly served the welfare of the many by producing a surplus far greater than the elite could enjoy. Here, much in the spirit of John Locke, Smith conjured up an image of a large estate, where a landlord commanded his tenants to produce an abundant harvest. But the lord could only consume so much of the product: "The capacity of his stomach bears no proportion to the immensity of his desires."[64] From this "oeconomy of greatness" flowed a surplus that provided "necessaries of life" for thousands.[65] Thanks to the operation of this "invisible hand," the elite served the interests of society "without intending it" or even knowing it.[66]

Smith balanced this critique of the rich with a more sympathetic account of prudence and work among the common people. The "desire of bettering our condition," he observed, "comes with us from the womb, and never leaves us till we go into the grave."[67] All humans shared an impulse to work assiduously and to "save and accumulate" the fruits of their labor. In a liberal country with secure property and free enterprise, this impulse found ample reward. For "the greater part of men," the virtue of "frugality seems not only to predominate, but to predominate very greatly."[68] Fear rather than vanity was the motivating force for prudent people. They viewed bankruptcy as "the greatest and most humiliating calamity" that could "befal an innocent man."[69] This cautious approach also inspired a distinctive pattern of consumption. Prudent people purchased "durable commodities" of good quality and lasting value, buying land, buildings, and furniture, instead of spending on the wasteful and shortsighted consumption he associated with the elite.[70] Smith thus differed from Mandeville and Hume, for whom any kind of consumption benefited the economy by putting the wheels of commerce in motion. The cumulative effects of the spirit of accumulation and prudent consumption could be seen everywhere in England. Over the last hundred years, the "private frugality and good conduct of individuals" had produced a growing capital to cultivate the land, expand manufactures, and maintain an increasing population. Prudential accumulation and consumption thus went hand in hand with the improvement of the natural world. For Smith, it was this "universal, continual and uninterrupted effort" that laid the true foundation of national wealth.[71] Smith's version of Enlightened Scarcity thus viewed nature and the economy, not as opposites locked into an adversarial relationship, but rather as partners in a process of improvement. Scarcity would not be eradicated, but both nature and economy would develop in such ways that human life would become more tolerable, across the social hierarchy.

Here we come full circle back to agriculture. Smith's optimism about human striving ultimately flowed from his view of the natural world. In difficult times, the poor had to act with foresight and temperance to withstand temporary shortages—rationing what little grain they had left and tightening their belts. But according to Smith, nature was never at fault when the perpetual condition of scarcity turned into serious hardship. The real cause of human suffering lay in politics rather than nature. Smith's views on dearth and famine reflected the fortunate circumstances of the English economy in the eighteenth century. Apart from some remote pockets, England had not experienced a widespread famine since the sixteenth century. Lowland Scotland too had escaped famine after the 1690s. However, a pattern of recurring famine persisted elsewhere in Europe and in the British colonies. Ireland saw pervasive excess mortality after the bitter winter of 1740. To explain these variations, Smith drew a strong distinction between natural dearth and man-made famine. Dearth should be expected, he insisted, in all countries in the temperate, grain-producing climate zone. It was a recurring and eternal part of the natural order. But famine was a different matter. Smith categorically denied that harvest failure might cause a famine in the wheat-growing regions of Europe under conditions of free trade with a good transportation system. If all countries joined together into a free-trading union with easy communications by land or water, then "the scarcity of any one country" could always be remedied through imports from another part of the continent. Dearth became famine only when governments meddled with free trade in grain: "famine has never arisen from any other cause but the violence of government attempting . . . to remedy the inconveniencies of a dearth."[72]

This Enlightenment faith in the providential bounty of the natural world and the wisdom of the market also profoundly colored Smith's conception of the future. Since land was the most fundamental form of wealth, Smith reserved his greatest hopes for the British colonies in North America. There, land was plentiful and labor scarce. Consequently wages were high and couples married young. Population doubled every "twenty or five-and-twenty years."[73] The colonists also enjoyed secure property rights and light taxes. Smith had very little to say about the place of slavery in this order or the eradication of indigenous peoples by settlers. Extrapolating from these factors, Smith proposed that the American future was so bright that the New World would one day eclipse the power of Great Britain, a sentiment with which Hume also agreed. Political gravity within the British Empire would shift to the New World in "little more than a century."[74] As it turns out, Smith was not far off the mark with his prediction. But why was he so certain that Great Britain must decline in comparison with North America? The answer lies once again in his appeal to land as the ultimate foundation of wealth. In the Old World, where nations had long been settled with dense populations, the opportunities for growth were limited and the future tended toward a "stationary" condition. This occurred when a nation had "acquired that full complement of riches which the nature of its soil and climate . . . and its situation with respect to other countries allowed it to acquire." As a nation approached the limits to investment set by its soil and the climate, "the wages of labor and the profits of stock would probably be very low." Here, we glimpse again the agricultural foundation of Enlightened Scarcity in Adam Smith. The interplay of desire and markets ultimately depended on the fertility of land for the prospects of growth.[75]

Conclusion

At the end of the eighteenth century, a series of seismic shifts shook the social world that had shaped Hume and Smith's thought. The old regime teetered in France, giving way to constitutional democracy and then a radical republic. In Britain, the traditional order persisted, although in a climate of sharpening polarization. Edmund Burke denounced the French revolutionaries in a 1790 polemic that helped give birth to modern conservative ideology. Thomas Paine counterattacked with a defense of universal suffrage in *The Rights of Man*. At the same time, Mary Wollstonecraft explored the possibility of new spaces and roles for women in *Vindication of the Rights of Woman* (1792).

In the pressure cooker of radical politics, the Enlightened idea of scarcity found a new expression in the work of the philosopher William Godwin (1756–1836). Educated at the Hoxton Academy, one of the so-called Dissenting academies established for Protestants unwilling to subscribe to the articles of the Anglican Church, Godwin arrived at a political position profoundly hostile to the traditional order. He viewed the growing inequality of British society with horror and embraced a vision of a fundamentally egalitarian future society. Like Hume and Smith before him, he saw incremental improvement as the proper goal of political economy. He also agreed that the moral striving of the individual could shape and refine desire. Indeed, Godwin was even more ambitious in his vision for human improvement than Hume and Smith had been. Godwin's aim was nothing less than the creation of a new kind of society, where the desire for material refinement and social distinction would come to an end.

As Godwin carried forward the legacy of Hume and Smith, he reformulated it in a number of distinctive ways. For Godwin, human needs were limited in scope, confined to "food, clothing and shelter."[76] Simple food was the key to well-being: "A frugal diet will contribute infinitely more to health, to a clear understanding, to chearful spirits, and even to the gratification of the appetites."[77] Hence, simplicity and frugality were inherently pleasant and attractive in accordance with human nature. The multiplication of wants in commercial society merely distracted from genuine welfare. Where Smith had praised the division of labor as a spur to industry and wealth, Godwin thought it would be better to simplify society, rejecting "unnecessary employments" and abolishing the "manufacture of trinkets and luxuries."[78] "Our only true felicity," Godwin insisted, lay in the "expansion of our intellectual powers, the knowledge of truth and the practice of virtue."[79] If the incessant pursuit of accumulation could be abolished, people would acquire the leisure to focus on the true source of happiness: the perpetual improvement of the mind. This new stage also marked the beginning of genuine sympathy with all other human beings.

Curiously, Godwin thought the perpetual improvement of the mind might eventually lead to the discovery of earthly immortality.

Godwin's hopes for a just society rested on what he called the "equalization of property."[80] No person should accumulate more than necessary to fulfill the basic needs of food, clothing and shelter. If people by chance fell short of this minimum, they had the right to expect help from their peers. In a state of equal property, humans would become accustomed to share their surplus freely "to supply the wants of their neighbor."[81] This interpretation of social harmony was underwritten by the abundance of the natural world. Godwin argued that inequality and monopoly had led to a gross neglect of agricultural potential. If the land was properly cultivated, Europe might maintain "five times her present number of inhabitants."[82] At the same time, the amount of labor needed to cultivate the land was far smaller than commonly assumed: "not more than one twentieth of the inhabitants of England are employed seriously and substantially in the labors of agriculture."[83] Echoing Thomas More's concept of Utopian Scarcity from the sixteenth century, Godwin argued that once work was apportioned equitably among all people, each individual would have to labor only thirty minutes per day to supply "the whole with necessaries."[84] This was a full five-and-a-half hours less than the workload of More's Utopians.

According to Godwin, this new social order would emerge spontaneously through rational deliberation and voluntary acceptance. Borrowing Smith's term, Godwin thought people would develop a sense of conscience, an "impartial spectator," by which to assess the moral significance of their actions. Such an inner compass would awaken them to the frivolity of luxury and the need to apportion wealth according to need rather than want. For this reason, Godwin opposed the path of violent revolution and state coercion so prevalent at the time in France: true justice could not be imposed from the top down.[85]

With Godwin, we reach the outer limits of the Enlightenment conception of scarcity. While both Hume and Smith believed in the progress of the sentiments, Godwin insisted that people's minds were capable of such extreme advancement that they would simply transcend the impulse to consume beyond need. In combination with nature's fecundity, a humanity cured of its false desires would be able to live in blissful sufficiency. Godwin's vision of the future relationship between the economy and nature brushed up against radical new possibilities. Where Hume and Smith placed the prospects for improvement within the practical constraints of commercial society and its agrarian foundation, Godwin reformulated Enlightenment improvement to embrace a world utterly different from his own. In the next three chapters we will explore three major Finitarian responses to the Enlightenment version of scarcity in Romanticism, Malthusianism, and Socialism.

4

ROMANTIC SCARCITY

n December 1799, Dorothy Wordsworth (1771–1855) and William Wordsworth (1770–1850) moved into a modest house in the village of Grasmere
at the heart of the English Lake District. Their move coincided with the
first great age of tourism in the region. Well-to-do British travelers came in
flocks to enjoy the dramatic scenery of the mountains and lakes. The
Wordsworths shared this aesthetic impulse, but turned it toward lofty new
goals. Dorothy and William had been born in the nearby market town of
Cockermouth. They were the children of John Wordsworth, solicitor to the
grandee James Lowther, first Earl of Lonsdale. Not quite locals but also
certainly not tourists, the Wordsworths embraced the rural life in Grasmere as a source of inner renewal and spiritual transformation. By immers-
ing themselves in the social life and natural world of this small place, they
hoped to achieve a profound connection with the earth itself.

Their time in the village was lovingly recorded in Dorothy's journals.
Weaving together high and low, she wrote of friendship and toothaches,
gardening and insomnia, the work of the villagers and the cycles of the
seasons. For Dorothy, journal keeping, no less than romantic poetry, made
possible a new way of being in the world. Out of the daily routine of

household chores, nature walks, and conversations with neighbors and friends, she fashioned a life devoted to material simplicity and poetic experience. Consider, for example, the entry for June 20, 1802. After spending their late Sunday morning in the orchard, the siblings followed a favorite path out of the village while discussing household finances. Soon economic concerns were set aside. "We lay down upon the sloping turf. Earth & sky were so lovely that they melted our very hearts. The sky to the north was of a chastened yet rich yellow fading into pale blue & streaked & scattered over with steady islands of purple melting away into shades of pink."[1] For Dorothy, such encounters with the natural world had a restorative effect on the spirit, charging everyday life with poetic beauty and intense joy. Nature had the power to reorient the desires, away from the consumption of material goods and the striving for social distinction. Nature was not merely a source of resources to extract but a home, shared with many other species of animals and plants, to respect and cherish. By training the senses and the mind on the physical world, the observer could transcend the ordinary self, treading a path first opened by ancient mystics and philosophers. Dorothy wrote of the moment on the hillside: "It made my heart feel almost like a vision."[2]

On full display in Dorothy's journal and William's poetry is a romantic understanding of the relation between nature and economy. Not by accident, Dorothy wrote again and again in her journal of the comfort of circumscribed spaces. The vale of Grasmere was a sheltered microcosm, protected from the outside world. William, too, expressed this sentiment in his poetry: "Embrace me then, ye Hills, and close me in."[3] Mountains had become objects of beauty to the educated public during the Enlightenment. Crucially, the Wordsworth siblings went beyond mere aesthetic appreciation to celebrate the people and economy of the uplands. The mountains and marginal soils of the Lake District bred a special kind of virtue. For William, the landscape molded the psychology and morals of the inhabitants. While David Hume saw moral sentiment emerging in the commercial hustle and bustle of the city, the Wordsworths found virtue in humanity's engagement with nature. In the poem "Michael," William Wordsworth depicted the self-reliance and perseverance of a local shepherd as traits implanted by the difficult environment: "The common air, the hills . . . impress'd so many incidents upon his mind, of hardship, skill, or courage, joy or fear."[4] Lakelanders grew accustomed to a life of material simplicity and independence, far away from urban society and aristocratic fashion. Dorothy admired the self-sufficiency and small scale of village life. Even the pages of her journal were recycled, with the price of paper so dear.[5]

The Charms of the Countryside

This embrace of village life was part of a broader revolution in sensibility that swept Europe's middling sorts in the late eighteenth century. Instead of understanding scarcity as an incentive to improvement and commerce, a new generation of poets and philosophers believed that scarcity demanded material simplicity. Instead of validating desires and consumption as pathways to human happiness, they prioritized living within the limits of nature as the necessary foundation of virtue and true community. Although this Romantic notion of scarcity celebrated traditional notions of

James Baker Pyne, *Grasmere from Loughrigg,* **1859.** By the middle of the nineteenth century, Wordsworth's romantic experiment in simple living had become an object of middle-class tourism. *Credit:* Hanna Holborn Gray Special Collections Research Center, The University of Chicago.

restraint and limits, it departed from the Neo-Aristotelian and Utopian ideals by jettisoning conventional Christian morality in favor of a novel spirituality of nature. Inspiration for this alternative conception of scarcity came in great part from the philosophical writings of Jean-Jacques Rousseau (1712–1778).

Of all the eminent thinkers of the Enlightenment, Rousseau was possibly the most contrarian figure. A Genevan citizen by birth, from a modest background, Rousseau dazzled Europe with his learning even though he never received a formal education. He made contributions to political economy, political theory, and pedagogy while also penning two autobiographies. Like Adam Smith, he was enamored with natural history and promoted the botanical method of Carl Linnaeus. Though Rousseau lived in the public eye and became friendly with luminaries including Denis Diderot and David Hume, he remained deeply troubled by his own celebrity and longed all his life for solitude and an escape from commercial society.

In the 1750s, Rousseau shocked his Enlightenment contemporaries by mounting a frontal assault on the conventional understanding of civilization and progress. Life in the natural state, he argued in *The Discourse on the Origin and Foundations of Inequality Among Men* (1755), was the best possible condition for all people. The key to the good life was self-sufficiency: "So long as they applied themselves only to tasks a single individual could perform, and to arts that did not require the collaboration of several hands, they lived free, healthy, good and happy as far as they could by their Nature be."[6] In the absence of a division of labor and the institution of private property, contentment was within easy reach. Desires did not "exceed . . . Physical needs." For Rousseau, the faculty of human understanding was inextricably bound up with the state of the passions and the imagination. Since natural man had no knowledge of the world or the future, he had no reason to yearn for new things: "His imagination depicts nothing to him." The condition of humans in the natural state was insular and self-sufficient.[7]

Yet such harmony could not last. The drive for self-preservation among humans led them gradually toward a new state of being. Natural forces of different kinds—from small obstacles to wholesale disasters—provoked creativity and consciousness. By responding to external pressures of various kinds, natural men learned how to master nature, little by little. This new sense of control in turn "aroused the first movement of pride."[8] Early people formed families, learned how to use tools and build huts, introducing the earliest "sort of property."[9] According to Rousseau, natural men acquired "several sorts of conveniences unknown to their Fathers."[10] Soon, these desires became habitual and "degenerated into true needs."[11] From the proliferation of artificial needs followed discord and vanity. "Everyone began to look at everyone else and to wish to be looked at himself."[12] In this way, the march of progress led further and further away from the original equality. Rousseau argued: "iron and wheat . . . civilized men, and ruined Mankind."[13] Not only did improvement increase inequality, it also obscured the true origin of freedom. Civilized men, like domesticated horses, had come to love the shackles of their captivity: "They call the most miserable servitude peace," much like the barbarians who had given up their freedom in exchange for Roman baths and granaries.[14]

Rousseau staunchly opposed the notion, embraced by Hobbes, Barbon, and others, that the human mind was, first and foremost, governed by self-love. He argued that his fellow philosophers had made a cardinal mistake by failing to recognize that the selfishness of modern man was a product of particular social arrangements. When philosophers limited their inquiries to the social world wrought by private property, money, and commerce, they ended up with a blinkered view of human potential. To discover the actual tendencies of human nature, one had to strip it of all the trappings of modern life. This was the purpose of Rousseau's conjectural history of the "savage" stage.[15]

Natural man, Rousseau insisted, was indeed defined by self-love, but of a kind very different from that assumed by earlier philosophers. The object of what he called *amour de soi* was "our preservation and our well-being."[16] *Amour de soi* was "contented when our true needs are satisfied." Such needs were always limited in numbers and scope; they remained concrete and specific.[17] Self-love as Rousseau defined it was accompanied by another natural sentiment; pity operated in every individual by moderating self-love and, as such, provided the foundation for all the social virtues. "Indeed," Rousseau asked rhetorically, "what are generosity, Clemency, Humanity, if not Pity?" Even "benevolence and friendship" were grounded in pity.[18] Together, *amour de soi* and pity produced harmonious relations between people and between humanity and nature. Once humanity embarked on its ceaseless quest for ever more property, and it became important to people to display their riches, the "gentle voice" of *amour de soi* was drowned out by a different, louder, and more aggressive self-love which Rousseau termed *amour propre*.[19] Not unlike Nicholas Barbon's infinite "wants of the mind," this was a pleasure that came from feeling superior to others. In Rousseau's words, "the ardent desire to raise one's relative fortune less out of genuine need than in order to place one-self above others, instills in all men a dark inclination to harm one another, a secret jealousy." The result was "always the hidden desire to profit at an-other's expense."[20] The ability to feel pity and sympathize with other people had now been transformed into *identification*, the act of seeing one-self through the eyes of others.[21]

Modern man's psychological disposition sparked a new condition of scarcity. Whereas for natural man "desires do not exceed his Physical needs," people living in commercial societies were oppressed by a "multi-tude of new needs."[22] Their constant striving for more material riches made them lose touch with their inner self and corrupted their relation-ship to both nature and humanity. Rousseau summed up the alternatives: "What makes man essentially good is to have few needs and to compare himself little to others; what makes him essentially wicked is to have many needs and to depend very much on opinion."[23] Once people fell under the spell of *amour propre*, they lost the capacity to see beyond or to check their "greedy, ambitious, and wicked" self-interest.[24] Instead, they internalized a desire for ever more consumption and embraced the fact that their lives would be defined by endless toil. They became like a trained horse, who "pa-tiently suffers whip and spur," while their former selves would have been more like the untamed steed, who "bristles its mane, stamps the ground with its hoof, and struggles impetuously at the very sight of the bit."[25] This version of scarcity was not class-based, as it had been for Gerrard Winstanley, leader of the seventeenth-century Digger movement. In Rous-seau's world, all people were trapped in a vice that kept on tightening as human wants expanded.

Rousseau's critique of civilization took the history of stages and progress so central to the Scottish Enlightenment and turned it upside-down: the greater the complexity and sophistication of social and economic development, the more humans sank into corruption and depravity. Still, even after the rise of the institution of property and the end of what

Rousseau called the savage state, he saw ways of avoiding moral failure. When Rousseau considered positive prescriptions for political reform, he tended to favor societies with a simple division of labor. If mankind was "made up exclusively of husbandmen, soldiers, hunters and shepherds," it would be "infinitely more beautiful than" a society "made up of Cooks, Poets, Printers, Silversmiths, Painters, and Musicians."[26] Nature had endowed people with the instincts "to feed, to perpetuate, and to defend" themselves.[27] Men could turn these simple instincts into virtues by guiding them with reason and managing them wisely. "The ancient Republics of Greece" had prohibited all occupations of a "quiet and sedentary" sort that corrupted the body and enervated the "vigor of the soul."[28] In Greece, the state "where virtue was purest and lasted the longest" was Sparta, the nation without philosophers.[29]

In modern times, remnants of such virtues still persisted in republican states and rural societies on the periphery of commercial civilization. Rousseau often praised the simple communities of the Swiss Alps, where he had spent his youth. The mountainous country near Neuchâtel was dotted by small farms, "each one of which constitutes the center of the lands which belong to it," and their inhabitants enjoyed "both the tranquility of a retreat and the sweetness of Society."[30] Every farmstead functioned as a self-contained unit: "each is everything for himself, no one is anything for another." The peasants were free, lived in comfort, and, unlike their French counterparts, were not subject to excessive taxes or forced labor. Swiss people exhibited "an amazing combination of delicacy and simplicity" that Rousseau had "never since observed elsewhere."[31]

For Rousseau, the self-sufficient habits and values of the Swiss served as an inspiration to imagine an alternative path of human flourishing— the condition we call Romantic Scarcity. In his sketches for the constitutions of Corsica (1764–1765) and Poland (1771–1772) he set out to explain how a nation might avoid the pitfalls of commercial society. In the Polish case, perhaps the greatest challenge to achieving this ideal was the sheer size of the country. For true patriotism and democracy to flourish, citizens must feel they are constantly in the public eye. "Almost all small States, republics and monarchies alike," Rousseau noted, "prosper by the sole fact that they are small, since all the citizens in them know each other and watch each other, since the leaders can see by themselves the evil that is done, the good they have to do; and since their orders are executed under their eyes."[32] A second critical factor was to limit the influence of money. By converting the army into a national militia along Swiss lines, the Polish government could avoid a huge financial expense. In this way, Rousseau hoped to resist not just the logic of capital accumulation but also the growth model embodied by military states funded by public debt and heavy taxes.

In the case of Corsica, Rousseau argued that geographic insularity and social simplicity would allow the country to follow the Swiss path. Mountains, islands, and a largely rural population helped insolate society from moral corruption. In the plan for a Corsican constitution, Rousseau resisted the use of money and long-distance trade. Taxes should be paid in kind and the size of administration kept to a minimum. Agricultural labor was the best occupation for the people, encouraging physical vigor and peace of mind. Whereas commercial polities like France and Britain inflamed the passions of their populations with objects of consumption that stirred up envy and competition, Rousseau's constitution would channel the desires of Corsica's citizens toward simple needs and relative equality in the austere spirit of Sparta or republican Rome. Farmers who cultivated the land were by nature more attached to the nation than cosmopolitan city-dwellers were. Since the demanding and diverse character of agricultural labor required "constant attention," it prevented rural people from developing the vices associated with leisure. Farming work made them "patient" and "robust" in spirit.[33]

Proper pedagogy provided another key to moral probity. Rousseau hoped to instill in Corsica's children the right norms and habits. Here he followed closely the precept laid out in his treatise *Émile* (1762): "observe nature and follow the path it maps out for you."[34] Rural people should be guided by the moral authority inscribed in the natural order. Agricultural work was the most "decent, the most useful, and consequently the most noble," though the artisanal trades, such as ironworking and woodworking, were also respectable and salubrious.[35] Manual labor generally brought the workers "closest to the state of nature."[36] Rousseau welcomed refinements in the arts or improvements in technology, not as a means to control the natural world in the sense of Bacon or Hartlib, but rather as a way to fulfill truly essential needs. Yet the defense of this constitution contained a fatal weakness: the internal harmony of Corsica required an agrarian economy too small and simple to protect the nation from any external aggression by richer neighbors. Rousseau never explained how his austere virtues could safeguard the independence of his new republic in an age of commercial warfare and imperial rivalries.[37]

While Rousseau's political visions failed to bear fruit, his ideal of Romantic Scarcity was easier to embrace in private life. Rousseau himself made clear in his autobiography that the peace and tranquility of the simple life was not reserved for local farmers but could also be experienced by educated people. In 1765, Rousseau spent two months on the island of St. Pierre in Lake Bienne, near Bern. He described the pleasure of solitude in ecstatic terms. On the island, he felt entirely "self-sufficient, like God."[38] Such autonomy was accompanied by a profound change in his perception of time. During his stay on the island, Rousseau felt no need to "recall the past or encroach upon the future." Instead, his sense of the present ran on without a sense of duration, indefinitely.[39] This experience closely resembled Rousseau's idea of early human life. For prehistoric people, the experience of time was closely tied with sharply bounded desires: "His modest needs are so ready to hand . . . that he can have neither foresight nor curiosity. . . . His soul, which nothing stirs, yields itself to the sole sentiment of its present existence, with no idea of the future, however near it may be, and his projects, as limited as his views, hardly extend to the close of day."[40]

The Simple Life

In the late Enlightenment, the dream of the simple life found a popular audience through works of fiction and poetry. Rousseau himself paved the way here with his novel *Julie, or the New Heloise* (1761). This was the story of the doomed romance between a young noblewoman and her middle-class tutor, told through a tempestuous exchange of letters. Although Julie acquiesced to an arranged marriage, the novel ended happily with husband, wife, and lover reunited in domestic harmony on Julie's estate in the Alps. Here they could follow the precepts of nature in a sheltered microcosm far from city life. Rousseau's book became wildly popular with eighteenth-century readers. Rustic manners and mountain scenery also added to the broad appeal of the narrative. Indeed, Rousseau defended the merits of his novel as a rare and singular work of literature that would induce virtue, as long as it was read at a great distance from Parisian high society. A generation later, Rousseau's student and friend Jacques-Henri

Bernardin de Saint-Pierre (1737–1814) reworked many of these themes in his bestselling 1788 novel, *Paul et Virginie*.[41] The protagonists of the title were two shipwrecked children growing up in Arcadian innocence on the island of Mauritius. Where their predecessor Robinson Crusoe had used his island solitude to remake himself into an agent of bourgeois industry, Paul and Virginie embraced a self-sufficient household economy that kept them safe from the artificial and vicious desires of urban society. They knew nothing of the past or the future beyond the bounds of their mountain: "Solitude, so far from making them savages, had made them more thoroughly civilized. If the scandal of society gave them nothing to talk about, nature was at hand to fill them with delight."[42] *Paul et Virginie* enjoyed popular success into the nineteenth century, though curiously its readership shifted from adults to children over time. Dorothy and William Wordsworth were both avid readers of Bernardin de Saint-Pierre. We can understand their move to Grasmere in December 1799 in part as an attempt to emulate the virtues and sentiments of *Paul et Virginie*. Here was a northern counterpart to the secluded island home in the novel. Dorothy and William were self-consciously embracing a simple, self-sufficient existence, purged of artificial desires, what Dorothy called "plain living but high thinking."[43]

The house at Grasmere had until recently served as a coaching inn, called The Dove and Olive Bough, on the main road between Ambleside and Keswick. There were four small rooms to each floor. Downstairs was a living room with dark wall panels, stone floors, and a cooking range. In the back was a buttery cooled by an underground streamlet. Upstairs, Dorothy papered the walls of the bedroom with newspapers to keep out the cold. The rooms were furnished comfortably but without ostentation. As a working household, it was also a simple operation. Dorothy had the help of a neighbor who did the cooking and washing. From the beginning, she and William saw their new home as a "cottage." This word had acquired a new, special ring in the eighteenth century. Improvers encouraged the building of functional cottages to house tenants on estates. Architects were also beginning to design genteel cottages for the wealthy as fashionable spaces of retreat from the city.[44] Dorothy made the idea her own by associating it with sibling love and the charms of a modest home. After the death of their parents, she had lived apart from William among relatives in different places. In a letter to a friend written in 1793, she imagined cottage life as a kind of earthly paradise: "I am alone; why are not you seated with me? And my dear William why is not he here also? . . . I have chosen a bank where I have room to spare for a resting place for each of you. I hear you point out a spot where, if we could erect a little cottage and call it *our own* we should be the happiest of human beings."[45] When Dorothy and William signed the lease for the house and renamed it "Dove Cottage," they were fulfilling Dorothy's dream of a safe haven and also beginning a self-conscious experiment in simple living, inspired by Rousseau and Bernardin de Saint-Pierre.

Life in Grasmere had a strongly communal dimension. Unlike Paul and Virginie, the Wordsworths had plenty of neighbors. Dorothy and William were both fascinated with the rugged character of local shepherds and farmers. William believed that the difficult terrain of the region expanded and elevated the mind by instilling virtues of endurance and self-sufficiency. Sheep farms were not idylls of pastoral repose but places of relentless and solitary labor. In the poem "Michael," Wordsworth told the story of an aging shepherd who sent his son away to pay off a debt to secure the patrimony of the farm. But he lost both farm and heir when the son fell in with bad company in the city. For Wordsworth. Michael's only error was that he loved the farm "even more than his own Blood."[46] This was not simply a matter of poetic sentiment to Wordsworth but a political observation of great significance. Wordsworth believed that the small farmers of the Lake District, known locally as "statesmen," presented a bulwark for British liberties against radicalism. Writing in a time of economic dearth and revolutionary turmoil, Wordsworth suggested that the independence and modest needs of his shepherd-farmers offered a moral example for poor people everywhere. This was the best remedy against servile dependence on "workhouses . . . and Soupshops."[47]

Amour propre in Rousseau's sense held little sway in Wordsworth's social order. In the poem "Michael," the shepherd and his wife live a simple life of few wants. Their diet consists of "pottage and skimm'd milk . . . with oaten cakes and . . . plain homemade cheese."[48] Despite this meager existence, Michael and Isabel are entirely content. "We have enough," the shepherd tells his wife.[49] Among their few possessions is an old lamp—"an aged utensil"—which shines in the window of their cottage every night, a sign of simple constancy.[50] There was no place in Wordsworth's poem for Nicholas Barbon's restless version of human psychology—perpetual longing spurred by the desire for absent objects. Michael's only regret is the loss of his son. The bonds of family and community form the true sources of satisfaction.[51]

A shadow of doubt has long lingered over Wordsworth's pronouncements about the Lakeland peasantry. Although his poetry has been immensely influential, its social vision remains contested. Skeptical observers see Wordsworth's notion of the statesman-farmer as the brainchild of a certain kind of conservative idealism. Such skepticism finds support in the social circumstances surrounding his work as a poet. For all of Wordsworth's sympathies with shepherds and farmers, he lived apart from them, a Cambridge-educated, middle-class man who found national fame and eventually became poet laureate of Great Britain. Though he was a passionate advocate of hill farming, he never fully grasped its meaning or nature. When the clergyman Hardwicke Rawnsley collected testimony about Wordsworth's life and reputation a generation after his death, he found that local people had few kind words for the poet. They remembered him as an aloof outsider and even disparaged his talents as a poet. To gain a better sense of the experience of rural life in the period, we might turn instead to Wordsworth's near contemporary, the Northamptonshire poet John Clare (1793–1864).[52]

Neglected by critics and readers until the twentieth century, Clare is now recognized as a leading figure in romantic literature. In his lifetime, Clare struggled to find recognition. In contrast with Wordsworth's origins, his were unequivocally plebeian. His father, Parker Clare, was a farm laborer and the illegitimate son of a schoolteacher. Lacking connections and patronage, John Clare received a brief and uneven education in the local school. At the age of thirteen he came by a copy of James Thomson's poem "The Seasons" that inspired him to try his own hand at poetry. In 1819 a local bookseller put him in touch with a London publisher, opening the door to his brief literary success as a "peasant poet." But his later writings met with public indifference. In his forties, Clare succumbed to mental illness. The contrast between Clare and Wordsworth is sharp. After the lean years in Dove Cottage, Wordsworth was able to move to the far larger establishment of Rydal Mount. Profiting from his fame and connections, he secured a lucrative post as Distributor of Stamps for Westmoreland in 1812. By the time Wordsworth became poet laureate in the 1840s, Clare was locked up in Northampton General Lunatic Asylum, where he spent the final twenty-three years of his life.

Clare's poetry was shaped above all by the social experience of enclosure. An Act of Parliament enclosed his native parish of Helpstone in 1807, setting off the kind of hardship and dislocation that Winstanley had captured almost two hundred years earlier. The old landscape of open fields and commons was destroyed. Villagers could no longer claim customary use rights to gather firewood and graze livestock on common land. Clare's poetry describes in vivid detail the social and environmental devastation wrought by the new regime of property rights. In poems like "Helpstone," "The Mores," and "Remembrances," he bore testimony to the lost world of his childhood when the land was still held in common. This, Clare insisted, had been an age of "Peace and Plenty . . . known to all."[53] The landscape before enclosure was a patchwork of woodlands, heaths, greens, and other forms of "waste"—rich with resources accessible to the entire local community. In his poetry, Clare resurrected this landscape, reminding the reader of its complex geography and social meaning. If you could name all these things and places, you could also make a claim to possess the landscape. In "Remembrances," Clare hinted at the myriad ways in which the child learned about the uses of common land through play and work. "When jumping time away on old cross berry way / And eating awes like sugar plumbs ere they had lost the may." Like More and Winstanley before him, Clare was an eyewitness to the ravages of agrarian capitalism and the cruel logic of Enclosure Scarcity. But Clare's poetic sensibility also set him apart. He distilled from the experience of enclosure a romantic vision of community and the natural world quite different from that of More and Winstanley.[54]

The disaster of enclosure had leveled Clare's childhood world and turned it into a "desert by the never weary plough."[55] A multifaceted landscape rich in material uses and social meaning had been denuded and simplified to make way for widespread improvement.

> The bawks and Eddings are no more
> The pastures too are gone
> The greens the Meadows and the moors
> Are all cut up and done
> There's scarce a greensward spot remains
> And scarce a single tree
> All naked are thy plains
> And yet they're dear to thee.[56]

In the poem "Helpstone," Clare contrasted true and false abundance. The "Peace and Plenty" of the commons benefited the whole community whereas the "accursed Wealth" of enclosure was the property only of a "few."[57] This judgment rested not just on the value of equality but also on an economy of sufficiency. For Clare, a cottage home represented stability, shelter and the comforts of the hearth. One of the few modest triumphs of his difficult life was the offer from Lord Milton in 1832 of a "most comfortable cottage" with "an acre of orchard and garden, inclusive of a common for two cows, with a meadow sufficient to produce fodder for the winter."[58] Yet in Clare's poetry, the economy of the household could not be separated from the commons. This was a plebeian version of Romantic Scarcity, defending the needs and livelihood of the common people. Over and over again in his writings, freedom and value emerged from the love of simple pleasures associated with communal life and the natural world. The social historian Jeanette Neeson confirms that common land conferred invisible earnings outside the market system. But she also observes that the abundance of the commons presupposed a particular conception of desire: "Commoners had little but they also wanted less."[59]

The act of enclosure produced physical hardship for peasant occupiers by destroying woodlands and pasture. Clare turned to the animal world to convey his sense of horror. Farmers and gamekeepers would string up moles and other vermin on their fences as a warning to all pests and other trespassers. Such policy brought to mind the systematic terror and destruction wrought by Napoleon's reign on its conquered subjects.

> Inclosure like a Buonaparte let not a thing remain
> It leveled every bush and tree and leveled every hill
> And hung the moles for traitors—though the brook is running
>   still
> It runs a naked brook cold and chill.[60]

We see here how deeply the social and the natural world grew intertwined in Clare's mind. The defense of village communities went hand in hand with a keen appreciation of the rural landscape before enclosure. In this way, social criticism became a bridge toward an extended sense of community beyond the human realm. Moles were people, too.

Once Clare started to think this way about wild things, his poetry took an unexpected direction. In a series of astonishing poems about the birds of the local landscape, Clare began to imagine what the human community looked like from the *outside*. Snipes, sand martins, fern owls, thrushes, and nightingales all made their homes in the woods around Clare's native village. They, too, formed communities in distinct landscapes (the concept of the habitat came into use around this time). Their nests were miniature dwellings, built to offer comfort and security. But their lives were shaded by constant fear of outside threats—above all, human trespassers. Clare knew intimately the destruction wrought by hunters and collectors. He had grown up climbing trees and plundering nests for pleasure.

Such a bird's-eye view, looking down at people from the treetops, collapsed all distinctions of property and class, showing humans only as an undifferentiated and predatory mass. The same shift in perspective also revealed the intrinsic value of the natural world beyond economic use. In the woodlands, Clare found a sense of peace and refuge from the strains of village life and literary ambitions. Birds were free from "meddling toil" and "artificial toys" and "mercenary spirit."[61] This joyful encounter with the wild went hand in hand with an ethos of restraint. Clare no longer plundered nests but was content to observe and record. His eyes opened to the value of natural obstacles to exploitation. Wetlands offered safety from nearby human population. "Boys thread the woods / To their remotest shades / But in these marshy flats these stagnant floods / Security pervades."[62] Here was an ecological reason to resist enclosure, distinct from the defense of common use rights. A landscape that had not yet been drained and cultivated could serve as a sanctuary for wild things. More than a generation before the first move toward systematic conservation in Britain—the 1869 Act for the Preservation of Sea-Birds—Clare's defense of the traditional landscape nudged him toward a deep and radical sympathy with the diversity of nonhuman life forms.

The Stationary State

Clare was not alone in turning to the natural world for solace and pleasure. John Stuart Mill (1806–1873), for example, is someone now remembered principally as a philosopher and political economist, but he was also a lifelong plant hunter and amateur botanist. The young Mill, driven to nervous breakdown by his father's harsh pedagogical regime, looked to the natural world for escape and distraction. One of his proudest achievements was the survey he made of the flora in his native Surrey—incidentally, also home to St George's Hill, where Winstanley and his Diggers protested the enclosures. Mill's private passion for plants also influenced his social and political vision. In later life, he became a defender of common access to landscapes of outstanding natural beauty. He founded the Common Preservations Society and the Land Tenure Reform Association. Like Clare, Mill came to see human activities as a threat to the natural world. When the Royal Horticultural Society introduced a prize contest for the best herbaria in Britain, Mill sounded the alarm in a letter to *The Gardener's Chronicle and Agricultural Gazette* that such a competition might trigger a scramble of amateur collectors, with devastating ecological consequences. "Already our rare plants are becoming scarcer every year," he warned. "You are, no doubt, aware how rapidly, for example, the rare Kentish Orchids are disappearing." The herbarium contest might encourage ignorant "dabblers" to uproot and destroy native flora across the country so that "the present year 1864 will be marked in our botanical annals as the date of the extinction of nearly all the rare species in our already so scanty flora."[63] Like Clare, Mill worried that human activities, even in the form of well-intentioned scientific efforts of inventory, was diminishing the diversity of wildlife. He compared the present threat to the native flora to the outright extermination campaigns carried out against predators in the past, which had brought the wolf, bear, and beaver to extinction in the nation. Together with Charles Darwin, Mill helped organize a petition to the Royal Horticultural Society to alter the rules of the contest. They emphasized that botanical extirpation was a direct consequence of agricultural improvement. Because of high land values and intensifying productivity, "many wild plants" had reached the point of being confined to a few or even to single localities, often of small extent."[64]

Viewing Mill's work through a botanical lens, we gain a new perspective on one of the most puzzling and famous aspects of his work: his discussion of the stationary state in *The Principles of Political Economy* (1848). In this short chapter toward the end of the book, which was heavily influenced by his long-term partner, Harriet Taylor, Mill warned that the "richest and most prosperous countries would very soon attain the stationary state" unless "improvements were made in the productive arts" and capital was poured into "the uncultivated or ill-cultivated regions of the earth." Like the political economist Thomas Robert Malthus (to be discussed in Chapter 5), Mill feared that the speed and scale of modern growth was carrying the advanced economies toward a permanent ceiling beyond which they could not pass: "all progress in wealth is but a postponement of this . . . each step in advance is an approach to it." The prospect of stagnation was no longer distant but "near enough to be fully in view . . . we are always on the verge of it."[65] For Mill, this crisis also threatened the diversity and wilderness of the natural world, with "every rood of land brought into cultivation . . . all quadrupeds or birds which are not domesticated for man's use exterminated as his rivals for food . . . and scarcely a place left where a wild shrub or flower could grow without being eradicated as a weed." A crowded, domesticated world without wild spaces would harm the human mind irreparably, since "solitude, in the sense of being often alone, is essential to any depth of meditation or of character." Embracing a position that anticipated the conservationists of the late nineteenth century, such as John Muir, Mill observed that "solitude in the presence of natural beauty and grandeur, is the cradle of thoughts and aspirations, which are not only good for the individual, but which society could ill do without."[66]

Yet, the moral lesson of this forecast also made possible an alternative ending to the history of capitalism. In Mill's version of Romantic Scarcity, humanity could embrace the possibility of the stationary state "long before" the physical limits on growth became pressing and severe.[67] Such a choice would permit people to transcend the brutality and ugliness of industrial society. "I am not charmed," Mill noted wryly, "with the ideal of life held out by those who think that the normal state of human beings is that of struggling to get on; that the trampling, crushing, elbowing, and treading on each other's heels, which form the existing type of social life, are the most desirable lot of human kind."[68] In reality, the industrial age was merely a passing phase—a necessary stage in civilization, to be sure, but not the crowning glory of human society. This stationary society would be free to redirect its fundamental creative urges in new directions: "There would be as much scope as ever for all kinds of mental culture, and moral and social progress; as much room for improving the Art of Living, and much more likelihood of its being improved, when minds ceased to be engrossed by the art of getting on."[69] Throughout the chapter, Mill characterized the problem as a universal choice of the "species" rather than the path of a single class or a nation. He also framed the value of the stationary state in terms of stewardship and the preservation of wildlife: "It is not good for man to be kept perforce at all times in the presence of his species."[70]

Even though Mill's book *Principles of Political Economy* represents a great synthesis of nineteenth-century economic thought, it is actually a curiously uneven reflection of British industrial society. There is little detailed commentary in it on the factory system and the industrial slums. When Mill launched the term "industrial revolution," he used it to describe the intensification of industriousness in early commercial societies rather than the coming of the factory age.[71] A large portion of the first part of the book was occupied with a comparative history of land tenure. Though Mill did not support radical land reform, he saw moral value in ownership of small farms. A claim to land instilled virtues of "prudence, temperance and self-control" in the peasant class.[72] Here, he echoed Harriet Martineau's ideas of self-improvement and foresight (we will discuss Martineau's views in Chapter 5). But just as important was Mill's devotion to William Wordsworth's poetry and his pilgrimage to the Lake District in 1831. He praised the Lakeland hill farmers as a vestige of the "yeomanry" of the Middle Ages.[73] In this and other ways, Mill tempered his analysis of political economy with an idea of Romantic Scarcity that illuminated the potential for moral virtue among the rural poor.

For some Victorians, Mill's "Art of Living" was not a distant prospect but a matter of urgent action. In 1872, the political economist and artist John Ruskin (1819–1900) moved from London to the shores of Coniston Water in the English Lake District. He came to the north in search of refuge. The countryside offered a sanctuary from the consumerism and pollution of the Victorian city. At his house, Brantwood, overlooking Coniston Water and the Old Man—the great hill to the west of the village—Ruskin launched a utopian movement against mass consumption on behalf of the "workmen and laborers of Great Britain." At the heart of the project was the revival of handicraft industry in the Lake District between 1880–1920.[74]

For Ruskin and his allies, the aim of their movement was to anticipate a post-industrial future. In the twenty-ninth letter of his *Fors Clavigera* serial (1873), Ruskin urged his followers to look forward to a "sweet spring-time" for "our children's children . . . when their coals are burnt out, and they begin to understand that coals are not the source of all power Divine and human."[75] Ruskin's prediction echoed the forecast made by William Stanley Jevons in *The Coal Question* of 1865. Jevons had calculated then that British coal consumption would soon face increasing costs of extraction. Not too far into the future, Britain would lose its status as a great manufacturing nation and become a post-carbon society. Ruskin's arts-and-crafts community in the Lake District sought to establish an alternative economy—no longer dependent on coal and steam—but founded on skilled work and communal bonds. This vision of artistic artisans engaged in joyous creative work looked to a highly idealized version of medieval history to imagine the end of industrial capitalism.[76]

During the 1870s, Ruskin became increasingly concerned with the environmental degradation caused by industrial capitalism. From his windows at Brantwood, Ruskin charted unsettling and unprecedented phenomena in the skies above the Lakeland hills. The prevailing winds from the southwest brought smoke from the nearby manufactures on the coast. On his annual trips to the Alps, he bore witness to a warming trend in the mountains. As early as 1863, Ruskin had noticed that the glaciers near Mont Blanc appeared to be in retreat. Ten years later, he concluded that a third of the ice sheet in the Alpine glaciers had vanished in less than a generation. From these uncanny observations, Ruskin concluded that the climate was undergoing a sinister change, what he later called "The Storm-Cloud of the Nineteenth Century." Already in the fifth *Fors Clavigera* letter in 1871, he warned about the planetary reach of atmospheric pollution: "You can vitiate the air by your manner of life, and of death, to any extent. You might easily vitiate it so as to bring such a pestilence on the globe as would end all of you."[77] The ever expanding appetite of consumers threatened to make the entire world into a coal mine or factory. At mid-century, John Stuart Mill had seen the fundamental environmental problem as one of preserving rural haunts and wildlife from the encroachments of agriculture and suburban sprawl. But for Ruskin in 1871, the destructive power of industrial capitalism had coalesced into a new kind of threat. It was now a planetary force capable of polluting the atmosphere, even to the point of changing the earth's climate. Little did he know what the future held in store.

Strangely, the remedy for the Storm Cloud lay in the realm of the mind. Men and women must be taught not to want useless things. Wise consumption demanded an education of desire. "Three fourths of the demands existing in the world are romantic; founded on visions, idealisms, hopes and affections," Ruskin suggested, "and the regulation of the purse is, in its essence, regulation of the imagination of the heart."[78] The aim of the arts-and-crafts movement was to encourage consumers to reorient their desires away from conventional middle class goods toward art, history, and natural beauty. By refining the faculty of aesthetic judgment and the acquiring natural knowledge of the world, one would find new and better objects of desire. Ruskin also encouraged a deeper understanding of production processes. Wise consumption required a critical grasp of the labor conditions and the nature of supply chains: "In all buying, consider, first, what condition of existence you cause in the producers of what you buy; secondly, whether the sum you have paid is just to the producer, and in due portion, lodged in his hands."[79] In the place of industrial capitalism, Ruskin promoted an artisanal ethic of work that went against the grain of conventional political economy. Workers should confine production only to those articles that were genuinely useful to the consumer. Instead of flooding the world with cheap and disposable commodities, the workman should concentrate on objects of durable design and artistic merit that truly served human need and welfare. The "intrinsic value" of an object lay in "the absolute power" it possessed to "support life." Ruskin meant by this a mixture of biological necessity and aesthetic beauty: "A sheaf of wheat of given quality and weight has in it a measurable power of sustaining the substance of the body; a cubic foot of pure air, a fixed power of sustaining its warmth; and a cluster of flowers of given beauty, a fixed power of enlivening or animating the senses and heart."[80] By this standard, most middle class objects of consumption fell short of genuine value.

Ruskin's movement was at the same time a philosophical and practical experiment. Choices about what to consume at the level of the household also shaped the nation and the natural world. Through the education of desire, the Ruskinians sought to redefine the relationship between economy and nature. In practice, they tried to demonstrate that the good life depended on skilled work and artful simplicity rather than conventional status and wealth. This impulse animated a revival of handicraft as well as new currents in architecture, education, and landscape design. Central to the movement was a form of social preservationism, dedicated simultaneously to protecting the environment and the customs of the Lake

District. In this way, Ruskin and his followers hoped to foster a self-conscious culture of sufficiency, steering a middle course between abundance and deprivation.

Conclusion

Perhaps above all, what united the line of romantic thinkers from Ruskin back to Rousseau was a sense of the spiritual importance of nature to human welfare. Their main contribution was to imagine ways of dwelling in the world that limited human use and made room for the flourishing of other species. Rejecting the engineering ambition of seventeenth-century Cornucopian ideology as well as the industrialism of the nineteenth century, romantic thinkers refused to see the world merely as a resource, a standing reserve available for human exploitation. Clare's bird poems took stock of humanity from an external point of view. Mill wanted to make room for the nonhuman in the world by limiting economic growth. Ruskin presciently understood the planetary threat posed by industrialization, anticipating twentieth-century concerns about the overloading of the atmosphere with pollution. At the same time, romantic thinkers spurned the restless play of consumer desire. To be at home on earth was to limit human wants and economic growth, choosing a simple and slow life open to the natural world. Romantic Scarcity thus weaves together a philosophy and an aesthetic of the organic interplay between human and nonhuman lifeforms.

In political terms, romanticism has left an ambiguous legacy. One current of the movement tended toward illiberal nationalism. The fascination with peasant life and self-sufficiency produced disturbing xenophobic and racist echoes in twentieth-century fascist ideology. To take one example (discussed further in Chapter 8), Martin Heidegger's existentialist philosophy of dwelling was tainted by his dalliance with National Socialism. It would be a serious mistake, however, to equate romanticism exclusively with antidemocratic forms of ideology. As we have seen, one of the roots of romantic thought began with Rousseau's republican projects. A similar radical and democratic current surfaced in Clare's defense of common use rights and Mill's post-materialist stationary state. New versions of subaltern and radical romanticism have flourished in different corners of modern environmentalism, including the movement for climate justice and degrowth within Planetary Boundaries.

5

MALTHUSIAN SCARCITY

I t was a summer fit for the apocalypse. Vicious storms lashed the coasts of Europe in 1816. Torrential rains flooded towns and villages from Amsterdam to Geneva. The dismal weather persisted into the fall with fierce cold, hailstorms, and abundant snow. When the Quaker naturalist Luke Howard tabulated his temperature observations at the end of the year, he was astonished to discover that the average daily temperature in London had fallen by 12 degrees, from 50 degrees Fahrenheit in 1807–1815 to 38 degrees in 1816. On his honeymoon at Weymouth bay, the landscape painter John Constable captured the claustrophobic onslaught of blackened skies. Held up at Lake Geneva by the dismal weather, Lord Byron watched an ash cloud blot out the light of day. In a poem simply entitled "Darkness," he imagined the collapse of human society after the death of the Sun. Another romantic poet, Samuel Coleridge, joked in a letter about the "end of the World Weather." The cause of the preternatural darkness, unknown to all observers at the time, was the April 1815 eruption of the volcano Tambora in the Dutch East Indies (present-day Indonesia). This massive explosion halfway across the world unleashed a torrent of dust, ash, and other particles into the higher atmosphere. Over the course of the next eighteen months, these aerosols cooled global temperatures and shifted precipitation patterns to culminate in the infamous "Year without a Summer."[1]

Shortages went hand in hand with the Tambora eruption. The harvest was frighteningly late in the summer of 1816. Across northwestern Europe, crop yields fell by 75 percent. Ireland seems to have suffered a full-scale famine. Such brutal facts were not lost on political economists. David Ricardo (1772–1823) and Thomas Robert Malthus (1766–1834) exchanged unsettling observations about the dreadful weather and the plight of the poor. The historian, political economist, and utilitarian philosopher James Mill (1773–1836), father of John Stuart Mill, predicted to Ricardo that the "perfect continuance of rain and cold" would trigger a famine in which "one third of the people must die." Weighing present pleasure against future pain, Mill suggested that these people might be better off dead sooner rather than later. "It would be a blessing," he insisted, if the poor could be taken "into the streets and high ways," and have their throats cut "as we do with pigs." Otherwise a "whole people" would have to be fed by charity.[2]

While few contemporaries endorsed James Mill's brand of poor relief, his comments were emblematic of a new, bleaker tone in political economy. Concerns with material shortages, overpopulation, and the physical limits to growth became urgent in British economic thought at the turn of the eighteenth century. This change was as much social as environmental. Agrarian and demographic pressures went hand in hand with new political priorities and social values. Anxieties about the imbalance between population growth and the grain supply gained force in a highly peculiar social and political situation, marked by protracted warfare and commercial disruption, as well as ideological confrontation with the revolutionary regime in France. As the Revolution gathered pace on the continent, the view of the poor darkened in Britain. The resulting social and political tensions in turn entered into the interpretation of nature in political economy. Disruptions to the food supply came to seem far more threatening than before. Dearth was becoming a scourge of the social order.

Under these manifold pressures, the Enlightenment model of scarcity articulated by Hume and Smith began to lose ground to a new, more pessimistic strain of political economy. Malthus and his followers warned about the physical limits to growth imposed by the finite supply of land and nonrenewable mineral stock. Whereas Hume and Smith had devoted their attention to the benign effects of the passions on society, the Malthusians feared that sexual desire had the power to undo every effort at improvement. Smith had looked with compassion and admiration at the prudence and industry of working people, while Malthus saw them as creatures in thrall to base urges, incapable of rational foresight. The moral failings of the common people explained the persistence of misery. The poor failed to exercise the preventive check that curbed sexual desire by delaying marriage and reproduction. This dark picture of poverty in turn opened the door to new kinds of cruelty later in the nineteenth century. In the most infamous case, the British government denied aid to starving people in Ireland and India because officials thought famine relief would merely exacerbate the problem of overpopulation.

Malthusian Scarcity was born out of a specific historical moment of crisis, yet as an intellectual force it has proven immensely persistent and adaptable over time. This chapter traces the arc of Malthusian Scarcity from its origin in the troubled 1790s to its popular apogee in the 1840s, exploring why and how the idea became a staple of middle-class society. By tracking the origin and spread of Malthusian Scarcity, we also come to understand better why many critics have come to loathe this particular interpretation of nature and economy. Side by side with Malthusian political economy, the Romantic movement developed a polar-opposite conception of the natural world as a place of spiritual renewal and sufficiency. Meanwhile, the Socialists regarded the idea of Malthusian limits as an ideological weapon wielded by the elite to oppress the poor. Later in the century, neoclassical economists articulated their own critique of Malthusian Scarcity. The very idea of natural limits came to seem a childish fallacy to the advocates of infinite growth.

Against Granaries

To understand why Malthus was so pessimistic, we must begin by grappling with the social and political context of this thought. British farmers suffered a string of bad weather in the 1790s. Drought damaged the harvest of 1794, while the following year brought subzero winter temperatures, springtime floods, and then an unusually cold summer. After a brief reprieve, the difficult weather returned with cold and rain in 1799, followed by drought again in 1800. The prospect of dearth provoked widespread alarm and put political principles to the test. Observers disagreed about whether the government should intervene or stay aloof in the crisis. There were also concerns about how the common people might react in the face of severe dearth. The ancient dread of two consecutive bad harvests was still very much alive. Prime Minister William Pitt had been a follower of Adam Smith since the 1780s, yet his government had to steer a careful course between free trade and a more pragmatic commitment to social stability.[3]

The price of wheat nearly doubled to 90 shillings per quarter in 1795. In many localities, people worried that merchants might buy up the whole stock of grain and corner the market. There was some merit to this fear, since the nation's major cities exercised an increasing pull on agricultural production. Wholesale merchants moved up and down the country in search of wheat. By the middle of the summer, popular protests erupted across central England, from the Welsh borders to East Anglia. Rioters executed forced sales of wheat at prices they deemed more just than the going rate, with the proceeds accruing to the dispossessed corn merchants. Roads were cut off and ports blockaded to prevent the export of grain to metropolitan areas. Miners in the Forest of Dean and Cornwall also took part. A statement of the Cornish miners' position declared they would "not bear Starving when they see Grain carried out of the County without any brought to Market where every-one may see there is Corn & may Purchase at a Price Demanded."[4] Local magistrates often gave their support to such activities. But however justified in local terms, this kind of moral economy had deleterious consequences elsewhere, since it threatened to deprive the cities of the industrial Midlands of their basic food supply. Only the mobilization of the army and units of Volunteers broke the stalemate and reestablished the flow of grain from country to town.[5]

In the fall of 1795, King George III expressed his concern with the steep increase in the price of provisions. A Select Committee was formed to examine the matter. The Whig politician Charles James Fox suggested that the price of labor ought to be raised, although he was quick to add that this was a task for the landowners rather than the government. A more radical proposal was made by a friend of Fox's, the reformer and abolitionist Samuel Whitbread, who introduced into Parliament a bill to empower local magistrates to stipulate minimum wages according to the price of bread. Though Whitbread's plan was eventually rejected, it elicited an angry response by the aging statesman and political thinker Edmund Burke (1729–1797), published posthumously as *Thoughts and Details on Scarcity* (1800).[6]

Burke is usually thought of as one of the fathers of conservative thought, a defender of chivalry, monarchy, and the established Church against the schemes of dissenters, "oeconomists," and revolutionary radicals, but on questions of political economy his position in 1795 was uncompromisingly liberal. Against Whitbread, he argued that wage regulation was counterproductive. Even in a time of harvest failure like the present moment, the public must stick to liberal principles and reject government intervention. Whitbread's scheme to establish a minimum wage was nothing but a "discretionary tax on labor" according to Burke.[7] Labor was a commodity whose price should be set through free exchange: "The balance between consumption and production makes price. The market settles, and alone can settle, that price."[8] Burke defined "market" as the "meeting and conference of the consumer and the producer, when they mutually discover each other's wants." Every impartial observer, he insisted, must recognize the ability of the market to fulfill the "balance of wants" with "truth," "correctness," "celerity," and "general equity."[9] Any government interference with the "balance" of the market must lay an "axe to the root of production itself."[10] Attempts at wage regulation would simply diminish the productivity of farms. In fact, agriculture was particularly vulnerable to tampering, since cultivation demanded so much attention, skill, and capital by the landlord and farmer. A commercial system of agriculture could only operate efficiently if farmers were rewarded for their work with market-determined profits.

Burke's celebration of the market dynamic was not devoid of traditional principles. The "laws of commerce" were also the "laws of nature and consequently the laws of God." Providential design guaranteed the efficiency and justice of the market. Indeed, dearth itself was the choice of providence. The will of God shaped the pattern of feast and famine. "Divine Providence" was in charge, and when "it pleased" it to withhold "necessaries" from the poor "for a while," they must simply bear their fate. "Patience, labor, sobriety, frugality, and religion" would be required.[11] Government action could do nothing to soften acts of "divine displeasure."[12] Religion also played an important role in Burke's conception of social responsibility. If the wages of the workmen fell short of the price of provisions, their welfare came "within the jurisdiction of mercy."[13] Poor relief was not the responsibility of the state but fell squarely within the realm of private and voluntary initiatives. Burke added that "charity toward the poor" was a "direct and obligatory duty upon all Christians."[14] All of this was consonant with Burke's broader argument against revolutionaries and radicals: the Church was the anchor of both civil society and the state. Echoing the basic tenet of Neo-Aristotelian Scarcity, there could be no social order without organized religion.[15]

Like many other Enlightenment savants, Burke believed that providence operated through the orderly movements of the natural system. Dearth in agriculture was a recurring phenomenon, though it followed a pattern of "long cycles" rather than "short intervals."[16] Burke spoke from personal experience about farming. He owned the six-hundred-acre estate of Gregories near the market town of Beaconsfield in Buckinghamshire, where he witnessed firsthand the failure of the wheat harvest in 1795. Burke described in vivid detail the whole process, step by step. Frosts followed by rain damaged the early crops of both cereals and clover. When spring came, conditions appeared to improve. Grasses sown early revived and the wheat started blooming. Then, "at the most critical time of all, a cold dry east wind, attended with very sharp frosts . . . destroyed the flowers and withered up, in an astonishing manner, the whole side of the ear, next to the wind."[17] Burke showed some of the blighted wheat to his friends in town and warned about a bad crop ahead, but "his opinion was little regarded."[18] When the wheat was threshed, "he found the ears not filled, some of the capsules quite empty, and several others containing only withered hungry grain" resembling an inferior kind of rye rather than regular wheat. "Never had I a grain of so low quality," he lamented. Nevertheless, it sold at twenty-one pounds per load. Subsequently he sold two more loads for two pounds more each: "Such was the state of the market when I left home last Monday. Little remains in my barn."[19]

However disconcerting the present dearth was, one had to take a longer view. Like Adam Smith, Burke argued that famine was a thing of the past: "Never since I have known England, have I known more than a comparative scarcity."[20] While the historical record contained many instances of "melancholy havock," the price of wheat had been stable in recent years. There was, he added, good reason to believe that the common people fared better in "season of common plenty" than they had fifty years ago. Even the time of dearth was less hurtful than it once had been: "I do not know of one man, woman or child, that has perished from famine."[21] Here, Burke mixed into his political economy an explicitly political and moral theme by attributing the low rate of mortality to the strength of social bonds. The "care and superintendence of the poor" was "far greater than any I re-member."[22] Not unlike the Neo-Aristotelians of the sixteenth century, Burke considered paternalism in civil society to be the best remedy against dearth. Local elites rather than the central state had to shoulder this burden. Roman history demonstrated the danger of letting the gov-ernment assume too much responsibility over the food supply. The public granaries of Rome had encouraged dependency in the common people. "If once they are habituated to it, though but for one half-year, they will never be satisfied to have it otherwise."[23] The French monarchy had fallen into the same trap. Driven by noble but misguided intentions, the govern-ment there had sought to govern "too much," until the "hand of authority was seen in every thing and in every place"—encouraging people to blame it for everything that went "amiss in . . . domestic affairs."[24] Public rather than private paternalism bred corruption and insurrection. In the final in-stance, schemes to build granaries and regulate wages would provoke a revolution. A period of widespread if temporary shortages was far pref-erable to the siren call of state-sponsored famine relief. Short-term fluc-tuations in the grain price, for Burke, was a natural misfortune that the market could mitigate, with an occasional helping hand from Christian beneficence.

Island Limits

Just a year after Burke's death in 1797, the Anglican parson and political economist Thomas Robert Malthus launched a devastating attack on the idea that the human prospect was in harmony with the natural world. For Malthus, famine was not a problem of the past but a very real and unsettling threat to progress. The first edition of Malthus's *Essay on the Principle of Population,* published anonymously in 1798, targeted the vi-sion of human perfectibility articulated by William Godwin in *Enquiry Concerning Political Justice* (1793), discussed at the end of Chapter 4. No amount of economic improvement and social advancement could ever overcome the fundamental constraints to population growth posed by the finite supply of land. Public expenditure on the poor merely exacer-

bated the problem by encouraging early marriages and a surplus population of indigent and unemployed people. In the worst-case scenario, the
economy might degenerate into a stationary state, with population growth
pressing so much on agricultural production that the majority would be
forced to subsist in material misery.

While Malthus reached an extensive audience with his forecast of
overpopulation, he cut a far more modest figure than Burke's charismatic
and prolix statesman. After an early career in the Anglican Church, Malthus became professor of political economy at Haileybury's training college for East India Company officials. Though he did not own an estate, he
shared with Burke a deep affinity for rural England and the landowning
class. His thinking was also, like Burke's, profoundly indebted to the Scottish Enlightenment. But he combined his interest in the liberal political
economics of David Hume and Adam Smith with a keen appreciation of
Scottish writers on population such as Robert Wallace, Joseph Townsend,
and John Sinclair.

For Malthus, population growth was at the same time a natural and
an economic phenomenon. In this sense, political economy crossed into
the eighteenth-century science of natural history. Nature had scattered
the seeds of life liberally and profusely across "the animal and vegetable
kingdoms" but everywhere the available food supply set limits to the
growth of population.[25] This was a "prodigious waste of life," constantly
pitting living beings against each other in the "struggle for existence."[26] To
illustrate the exponential force of reproduction, Malthus proposed a
thought experiment. He asked his readers to imagine an alternative universe, where a species could expand indefinitely without competitors and
limits of subsistence. In such a counterfactual world, an animal or plant
would procreate and proliferate until it had colonized "millions of worlds
in the course of a few thousand years."[27] In the material sense, life was a
planetary force, capable of overwhelming and overflowing the universe.
This startling image was also a parable about human dominion. While
Homo sapiens was internally divided into warring states and empires, as a
species among other species, it lacked serious competitors. Thanks to the
art of agriculture, humans had learned to tame and transform the natural world. But there was a limit to mankind's mastery of the earth. The
global population could still overshoot food production in only a few generations: "In two centuries and a quarter the population would be to the

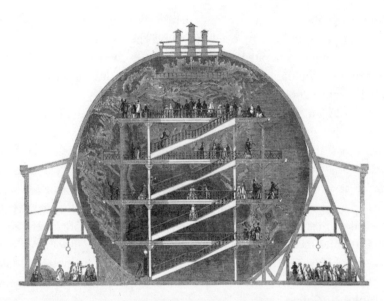

Wyld's Great Globe, Leicester Square, 1851. Images of the planet proliferated in Victorian popular culture, driving home Malthusian warnings about the physical limits to economic growth. *Credit: Illustrated London News,* June 7, 1851.

means of subsistence as 512 to 10."[28] This thought experiment encapsulated much of Malthus's method. To grasp population growth, one had to think on multiple scales, deploy long-range forecasts, and compare divergent rates of growth. Such thinking was increasingly commonplace among middle-class people at the time. Anxieties about population converged with new currents of scientific thought in geology and natural history. Images of the planet proliferated in popular culture, perhaps most memorably in the *Great Globe,* mapmaker James Wyld's sixty-foot model of the earth in London's Leicester Square.

Humans stood out from animals and plants by virtue of their powers of reason. They could choose to control their patterns of reproduction to fit their levels of income and social ambition. Unlike animals and plants, humans had the faculty of foresight and could elect to delay marriage and therefore reproduction to secure a sufficient livelihood before having offspring. But Malthus made it clear that poor people across the world frequently lacked the wherewithal and resources to provide adequately for their children. This peculiar definition of foresight suggested the unsettling possibility that some people were more like animals than others.

Among the poor in civilized nations, a "constant effort to increase population" frequently trumped the power of foresight.[29] Malthus also drew a line between civilized nations and morally degraded or barbarous ones. In the latter, the prudential check was weaker in force and early marriage more common. In the case of the British Isles, the worst afflicted area was the Scottish Highland region, where a demographic disaster was bound to happen since northern Scotland likely had the largest "redundant" population in the island.[30] The same tendency was even more pronounced in "morally degraded" nations, which for Malthus included many of the Catholic countries of Europe, in particular Ireland and Spain, as well as the Polynesian islands, New Holland (Australia), and Tierra del Fuego. In these places, people were so corrupted in their morals "as to propagate their species like brutes, totally regardless of consequences."[31] The ability to link cause and effect in reproductive terms was a defining feature of civilized and rational behavior. This emphasis on foresight was also crucial to Malthus's critique of the English Poor Laws. Since Elizabethan times, a system of parish-based poor relief had provided support for the indigent. Malthus believed that the Poor Laws inspired irrational behavior in the poor, including imprudent attitudes toward sex and early marriage. Only the wholesale abolition of the institution could instill proper foresight and self-sufficiency among the bottom ranks of society.[32]

Likewise in political economy, foresight was a crucial skill. Among Malthus's many arguments in the *Essay,* perhaps none is more striking than his forecast about population growth and agricultural improvement. Thanks to the abundance of land and the high wages of labor in the British colonies of North America, the population there doubled every twenty-five years. This notion of geometric population growth was something of a commonplace in the late Enlightenment; Malthus borrowed the observation from Richard Price, who in turn got it from Adam Smith, Ezra Styles, and Benjamin Franklin. In contrast with the American colonies, the British Isles had been settled a long time. Land was relatively scarce and the inhabitants many. Malthus speculated that the population of Britain could double within twenty-five years, but that any further expansion must be very difficult. The arithmetic development of food production simply could not keep up with geometric rates of growth in a long settled land. If population doubled again in a second twenty-five-year phase, calamity was inevitable. Redundant numbers must be eliminated by disease, war, or the "last, most dreadful resource of nature," famine.[33] While Adam Smith had seen mass hunger as an aberration brought on by bad governance, Malthus called it a providential check on excess population, an awful but natural result of the "prodigious waste of life" common to all of Creation.[34]

As if this forecast were not sufficiently gloomy, Malthus argued that population growth undermined the long-term prospects of every nation. China here functioned as a limit case. An ancient and very populous country, it had reached the full extent of agricultural productivity with "its soil nearly cultivated to the utmost."[35] Manufacturing and trade might bring more wealth to the few, but such development could not increase the basic "funds for maintaining labor." Only the "surplus produce of the cultivators" could guarantee a sufficient subsistence to the common people. When all land was fully cultivated, a commercial nation reached "the natural limit to the population." From this point onward, the "funds for maintaining labor" become "perfectly stationary."[36] In the long run, this was the fate of all commercial nations. Such projections would have a long afterlife in Victorian political economy (although not all of them were gloomy, as evidenced by John Stuart Mill's version of the stationary state, discussed in Chapter 4).

To anyone familiar with the history of British development, Malthus's worries might seem grossly misplaced. Why would any political economist be so anxious about the future just as Britain was entering the first Industrial Revolution? The simple answer is that Malthus, much like Adam Smith, did not grasp the promise of industrialization and the new fossil-fuel economy. Malthus's political economy was shaped by the intellectual inheritance of the preindustrial Scottish Enlightenment. It was also very much a reflection of the same experience of dearth, war, and revolution that formed Edmund Burke's economic thought. Malthus never quite managed to move beyond the moment of crisis in the 1790s, rather like a soldier who relives the battles of the past long after the war has ended. This is perhaps most obvious when we consider his view of the grain trade and agricultural production.

Malthus issued his warning at a moment when Great Britain appeared more isolated and vulnerable than it had been for centuries. The conflict with France underscored a danger easily forgotten in more fortunate times: when international trade became disrupted, the country had to rely on resources grown at home on relatively scarce arable land. This consciousness of island limits set Malthus apart from his predecessors in liberal political economy. Though he shared with Adam Smith and the French physiocrats an appreciation of the significance of agricultural improvement, he differed from them in his emphasis on the feebleness of technology and human ingenuity in the face of geographic restraints and demographic forces. Malthus's pessimistic assessment echoed wider anxieties about resource shortages among elites in the age of the French Revolutionary and Napoleonic Wars.[37]

Against Adam Smith's plea for free trade in grain across national borders, Malthus stressed the importance of government subsidies for exports. Since British wheat was more expensive than continental grain, subsidies were absolutely necessary to encourage national production. Without inducements, domestic farmers would worry about finding a market for their produce in years of good harvest and might be tempted to convert grain land into pasture to raise livestock. More meat production would benefit consumers in the middle classes but not the common people who depended principally on a diet of wheaten bread. Malthus made this argument against the background of a major shift in cereal production. Until the last quarter of the eighteenth century, Britain had been a net exporter of grain. But of late, this national surplus had given way to a pattern of import dependency. In a normal year, grain imports amounted to about 400,000 quarts. Even worse, the harvest failure of 1800–1801 had forced the government to import two million quarts from abroad at public expense. Given the pressure of population on available supply, Malthus predicted that such a dearth would be repeated in the near future. "We can hardly doubt," he wrote in 1803, "that in the course of some years we shall draw from [abroad] as much as two millions quarters of wheat, besides other corn, the support of above two millions of people." These supplies might come, he thought, from America and the nations around the Baltic Sea. But what would happen if there was a serious dispute with these countries in a time of harvest failure? Malthus added wryly: "with what a weight of power they would negotiate!" The entire Royal Navy of Great Britain would be less intimidating than the "simple threat of shutting all their ports."[38] Such a threat was particularly worrisome since Great Britain's "commercial ambition is peculiarly calculated to excite a general jealousy."[39] It would be foolish to trust in the goodwill of foreign nations under such circumstances. Even worse would be to sacrifice the lives of two million subjects of the Crown if no assistance could be found abroad. Much better, then, to minimize imports of grain and ensure national self-sufficiency; corn bounties would encourage British farmers to keep enough land under cultivation to ensure a buffer against years of poor harvests.

For Malthus, domestic grain production was crucial not just to national security and the welfare of the poor but to the survival of the liberal regime established in the Glorious Revolution of 1688–1689. Much like Burke, he linked the threat of revolutionary violence to the experience of material dearth in the lower orders. He suggested that the most dangerous threat to social stability was the "redundant" part of the population—that is, men and women who had little access to property and employment. If these people became convinced that the government was responsible for their plight, they would be easily seduced by radical promises and the lure of violence. In Malthus's phrase, "redundant population" turned into revolutionary "mobs" who were "goaded by resentment for real sufferings" but "totally ignorant of the quarter from which they originate."[40] Malthus suggested that Britain had come very close to disaster during the "late scarcities." Should such episodes become more frequent, something Malthus feared was all too likely, this might jeopardize the English Constitution. When "political discontents blended with the cries of hunger," the door was opened to revolution. Mob rule went hand in hand with dictatorship. Writing in the aftermath of Napoleon's rise to power, Malthus warned that "almost every revolution, after long and painful sacrifices, terminated in military despotism."[41]

Despite postwar depression and social unrest, the nightmare of a famine-driven British Revolution failed to materialize. When liberal critics attacked the system of agricultural protection at the end of the Napoleonic Wars, Malthus continued to defend the need for corn bounties. The peace between France and Britain did not, in his thinking, fundamentally alter the strategic and political necessity of securing the grain supply. Neither did Malthus make any major changes to his basic argument in the 1817 edition of the *Essay on Population*. He still warned that Britain would be unable to feed a rapidly growing population for long, and continued to insist that a serious dearth might provoke mob violence and a revolution.[42]

In the debate over the Corn Laws, Malthus's friend and sparring partner David Ricardo emerged as a formidable defender of free trade. In *The Principles of Political Economy and Taxation* (1817), Ricardo pushed the tradition of political economy toward a more deductive approach, away from the comparative historical analysis and moral philosophy of Hume, Smith, and Malthus. Ricardo's practical circumstances and social orientation differed a great deal from his predecessors. He was an enormously successful London financier who used his fortune to buy an estate and gain prominence in the English landowning elite. He was also a man of unconventional religious views, who converted from Judaism to join the Unitarian church, but may have been an agnostic at heart.

Much like Malthus's political economy, Ricardo's thought grew out of the emergency conditions of wartime Britain. One of his main contributions was a theory of rent that explored the economic effects of population growth on marginal soil cultivation. According to Ricardo, rent was the result of the expansion of cultivation. When less fertile soils were taken into cultivation, it became possible to charge rent on superior lands. Population growth therefore channeled wealth into the hands of the landlord class at the expense of more productive economics sectors. Yet while Ricardo was not oblivious to the problem of physical limits, he rejected Malthus' focus on self-sufficiency and protectionism. He hoped that the mutual advantages fostered by international trade could stave off the stationary state. In its simplest form, Ricardo's model took the form of a thought experiment about international trade and specialization, looking at the commerce between England and Portugal. He started with the observation that each country in his case could produce its own cloth and wine if it chose, but the two activities would demand of it different labor inputs reflective of the manufacturing capacities and real wages in those sectors, and the country's natural advantages. Assume, for example, that for England to produce an amount of cloth that would command a certain price in international trade, "the labor of 100 men for one year" would be required, but to produce an equally priced amount of wine it would have to devote the "labor of 120 men." Meanwhile, assume the same output would require of Portugal only "90 men" for the cloth and "80 men" for the wine. [43] Ricardo's insight was that, even though Portugal had a labor-cost advantage for both goods, it should specialize in the activity that would maximize its returns—winemaking—and, rather than expend its labor inputs in the less efficient business of textiles, import cloth from England. The same logic should drive England to specialize in cloth-making and to import wine, creating a tidy argument for bilateral trade. Ricardo's simple mathematics carried persuasive force, elegantly convey-ing the argument for specialization based on relative advantage. Taking aim at Malthus, Ricardo also suggested that grain production should be part of the international division of labor. If a country possessed "very considerable advantages in machinery and skill," it should "import a portion of the corn required for its consumption."[44] He praised John Ramsay McCulloch's proposal for a gradual liberalization of the grain trade. In a world of free trade, the "nations of the earth" would be "like provinces of the same kingdom."[45] McCulloch saw in international spe-cialization a defense against dearth. Consumers everywhere would draw the benefits of the most efficient production: grain from Poland, cotton from the United States, and manufactured goods from Birmingham and Glasgow. Such true commercial spirit would secure "permanently . . . the prosperity of nations."[46] "By the extension of foreign trade, or by improve-ments in machinery," Ricardo observed, "the food and necessaries of the laborer can be brought to market at a reduced price."[47] Free trade was thus an instrument to overcome the constraints of nature. Although Ricardo did not rule out the arrival of the stationary state, he hoped that it was "yet far distant."[48]

The concept of comparative advantage reduced a complex set of po-litical, social, and environmental variables into an ingenious but highly abstract model. As a moral and political proposition, Ricardo's notion of international specialization gave a new lease on life to cosmopolitan liberalism, with its idealist assumptions of mutual advantage and peace-ful exchange. But there were also a number of omissions from the model, which critics were quick to exploit. Like Malthus's original model of geometric and arithmetic growth, Ricardo's concept rested more on math-ematical intuition than on empirical data. The argument about specializa-tion could also be challenged at the level of its assumptions. How could one know the efficiency of Portuguese cloth-making without developing the industry there? What if developments in England allowed it to increase its yield in wine production? Questions like these cast doubt on the practical application of Ricardo's model. And beyond the challenges to his deduc-tive method were other, deeper questions: To what degree was the manufacturing strength of Britain the product of historical patterns of violence and exploitation? Did it matter that Portugal had long been part of a British sphere of economic and military influence? Skeptical readers of Ricardo, Karl Marx among them, concluded that Ricardo's notion of free trade was little more than an apology for unequal exchange.

Popular Malthusianism

To grasp the impact of these debates about Malthusian Scarcity on nineteenth-century culture and politics, we need to look beyond the pages of Malthus and Ricardo to the proselytizing efforts of their allies and friends. Among them were three Scots: Ricardo's student John Ramsay McCulloch, the first professor of political economy at the University of London from 1828; the Church of Scotland minister and social reformer Thomas Chalmers; and the Anti-Corn Law agitator James Wilson, who founded *The Economist*. But by far the most important figure of the moment was the English writer and journalist Harriet Martineau (1802–1876), who became an overnight sensation with the first installments of her *Illustrations of Political Economy*. Issued once a month in twenty-five volumes between 1832 and 1834, these were engaging tales that also taught lessons in economic theory. With her talent for combining entertainment and edification, Martineau quickly gained a devoted following which included the future queen of England and such intellectual luminaries as Robert Owen, and Thomas Carlyle. Each installment is believed to have sold ten thousand copies and by one computation may have reached as many as 140,000 readers in circulation. By comparison, the final part of Charles Dickens's first novel, *The Posthumous Papers of the Pickwick Club*, had 40,000 readers in 1837.[49]

Born into a Unitarian, middle-class family in Norwich, Martineau turned to writing professionally after the death of her father and the collapse of his manufacturing business. Though she had no more than a few years of formal education, her fierce intelligence and imaginative approach to economic reasoning made her a singularly effective advocate for classical economics. Much of the success of Martineau's project had to do with her ability to translate the sterile axioms of political economy into vivid narratives shaped by her Unitarian religious sensibility. She understood that middle-class Victorian readers would prefer their economic theory

enlivened by the idiom of literature and religion. One of her most celebrated novellas in *Illustrations of Political Economy* turned the problem
of Malthusian Scarcity into a story about how good housekeeping and foresight staved off disaster on a tiny island in the Hebrides.

While economic theorists aspired to state their principles in the most
universal form possible, Martineau used concrete and domestic settings to
stimulate public interest in economic arguments. She varied the social
context from story to story, but regardless of the class and rank of her protagonists, she always situated the action in local communities and familial
relations. By translating abstract ideas into dramatic plots, Martineau
hoped to make economic principles intelligible to all social classes. People
who had no inclination or ability to read the treatises of Smith or Malthus
might eagerly engage with their ideas in a different genre. Martineau here
followed in the footsteps of Hannah More and other Christian moralists
who had successfully grafted Christian messages onto the form of popular
fiction. Martineau was in fact an even more relentless moralist than her
evangelical predecessors. Each of Martineau's stories concluded with a
shamelessly didactic "Summary of Principles." By frequently choosing plebeian figures as mouthpieces, she honored Malthus's aspiration to make
economic laws intelligible to the common people. For Malthus, education
was a critical component of his project to improve the lives of the working
classes. In the second edition of his essay on population, he suggested that it
might be possible to alter the conduct of the common people by teaching
them "a few of the simplest principles of political economy" on the model of
Adam Smith's program of parish schools.[50] Yet despite Martineau's best intentions, she was far more successful in appealing to middle-class interest
than converting workers for her cause. Indeed, the publication of *Illustrations of Political Economy* coincided with an era of bourgeois ascendancy.
Martineau's literary breakthrough occurred in the years between the first
Reform Bill (1832) and the New Poor Law (1834), just as the middle class was
coming into its own as a political force. Like More's religious fiction, Martineau's work flourished in the heat of a specific moment. The next generation
would cast a cold eye on her pedagogical pretensions. While Martineau's
reputation did not endure, her works helped make the idea of Malthusian
Scarcity into a national force in politics and culture during the 1830s.

Among Martineau's many readers was the heir presumptive to
the throne of Britain. The young Victoria awaited each installment

with great excitement. Her middle-class literary taste should not surprise, given the distinctly bourgeois stamp she would put on the monarchy during her long reign. Victoria's favorite of the twenty-four tales was
"Ella of Garveloch," a story of family and improvement set on a tiny island in the Hebrides. Martineau followed up the success of this novella with
a second story, "Weal and Woe in Garveloch." In sharply drawn vignettes,
Martineau evoked the landscape and drama of island life. The insular location served two complementary purposes. Since the late eighteenth
century, the Highlands and Western Isles had become a popular tourist
destination and literary landscape. Martineau's story tapped into the romance of the Gaelic West while giving new life to the popular idea that
barren soils and a difficult climate bred a special virtue of hardy self-
reliance. At the same time, Garveloch also provided a natural laboratory
for economic reasoning, very much in the spirit of Malthus and Ricardo. The isolated setting and physical boundaries of the island made it
easier to isolate and describe the basic factors of economic growth and
demographic limits.[51]

In her two stories about Ella of Garveloch, Martineau approached
the question of improvement from both sides of the Malthusian-Ricardian
divide—first as a question of agriculture and rent, and then as a matter of
moral foresight and population pressure. At the center of the narrative was
a family of orphaned children led by the precocious heroine, Ella. Eager
to better herself and improve the family's cottage and fields, Ella needed
knowledge of economic principles to make good decisions. Martineau provided Ella with a disquisition on the origin of rent, using a friendly laird
(a landowner from the gentry) as mouthpiece. When population expanded
and farmers took more land and coastal waters into use, he explained, the
best soils and most advantageous fishing spots would begin to yield additional value in the form of rent. This was not an "arbitrary demand by the
landlord but a necessary consequence of the varying qualities of the soil."
Tenants paid rent to the landlord for the productive advantage of using
better soils. Rent was a "symptom" rather than a cause of wealth, just as
Ricardo would have it.[52] By investing the profits from fishing into soil improvement, Ella made her cottage holding prosper. She happily began to
pay rent to the landlord as a token of her own success. The story of Ella's
improvements concluded with the promise of an economically advantageous marriage with the young trader, Angus.

In the next story, "Weal and Woe in Garveloch," the action moved forward ten years. Ella was now the mother of nine children. Many others on the island had also opted to raise large families while times were good. This growth in population put pressure on the limited sources of subsistence. After a bad harvest, the poor began to experience hardship as growth in demand outpaced supply, causing prices to rise. For Martineau, this event was entirely predictable. On such a small island, there were no confounding factors to confuse the public. "The people of Garveloch might survey their little district at a glance, and calculate the supply of provision grown, and count the numbers to be fed by it." Any prudent observer could easily "discern" how to "proportion . . . labor and food." A "truly wise" person would also take into account the "probability of bad seasons" in calculating the burden on family resources.[53] But many people chose to ignore sound economic knowledge. Instead, they "abused the poor farmer" and accused him of taking advantage of the situation. "They were slow to perceive that it was themselves and not the farmer who had made the change."[54] By raising large families, these people had "caused the increase of demand and the consequent rise of price." Martineau explained their psychological state as a form of "giddy" short-term thinking, prioritizing sexual desire over future welfare. Encouraged by a season of prosperity, they had decided to put all their trust in "the sight of plenty around them." Inveterate optimists, "they now supposed that their island was enriched for ever."[55]

The character of Ella of Garveloch looked at wealth with a different conviction than the other villagers. While Ella, too, had raised a large family, her actions were guided by rational foresight. The memory of an episode of dearth in her childhood shaped her attitude to household provisioning. Whenever she could, she had put aside savings for the event of an emergency. After dearth struck and prices rose, rather than blaming the farmer, she scrimped and rationed her supply of grain, letting herself go hungry to protect the youngest members of the family. She even concealed the situation to her children so as to alleviate anxiety, taking only her oldest son into her confidence. Though Ella spoke with sympathy about the plight of the poor, she rejected their improvidence and failure to plan ahead: "we knew that such stormy seasons come from time to time; and yet we acted as if we were promised plenty for ever."[56] In Martineau's story, Ella acts as the guardian of family memory and therefore also the best witness to an uncertain future.

"Weal and Woe" was focused on the practical problems of coping with cyclical dearth, but Martineau also hinted at a long-term solution to Malthusian Scarcity: as long as men and women learned to act prudently, it was possible to live well even in a world of stark physical constraints. Toward the end of the novella, Martineau suggested that a change had come over the island in recent times. Fewer and fewer people succumbed to the positive check of war and disease. Marriage had become "less general" and took "place at a later age . . . among the middling classes." There was reason to hope that the common people would "soon" follow this "example."[57] By making Ella the arbiter of moral action, Martineau located the problem of population squarely in the household and turned female foresight into a critical driver of improvement. Family relations offered the key to understanding material dearth. The ultimate cause of deprivation was moral and spiritual. In every soul, a battle raged between sexual desire and rational foresight. How individual men and women weighed present urges against future hardship determined the fate of families, communities, and even the nation itself.[58]

Thanks to Martineau and the other apostles of Malthusian thought, political economy became a potent force in society and politics during the 1830s and 1840s. Evangelical thinkers like Thomas Chalmers forged a peculiar synthesis of Malthusian ideas and Christian beliefs. Evangelicals, unlike economists, "believed that the hidden hand held a rod."[59] Providence acted through general laws rather than special intervention. In this spirit, the 1834 amendment to the Elizabethan Poor Laws subjected the "idle poor" to a brutal workhouse regime. Providential Malthusian ideology did even more damage during the Great Hunger (*Gorta Mór*) in Ireland between 1845 and 1849. The chief administrator in charge, Assistant Secretary to the Treasury Charles Edward Trevelyan, expressed a great deal of ambivalence about the very idea of assisting the victims. He confessed (in a private letter) that he did not think government should supply food or increase "the productive powers of the land." The remedy for the "deep and inveterate root of social evil" of overpopulation lay outside the political realm. "I hope I am not guilty of irreverence in thinking that, this being altogether beyond the power of man, the cure has been applied by the direct stroke of an all-wise Providence in a manner as unexpected and unthought as it is likely to be effectual."[60] Trevelyan believed that potato cultivation had encouraged a culture of myopia among the Irish poor. Like the giddy fishermen in Martineau's Garveloch, the Irish lived from day to day without seeking to improve their means or planning for bad seasons. High potato yields fostered population growth without social refinement. Trevelyan's diagnosis associated these subsistence problems with a lower stage of history. "The Irish small holder lives in a state of isolation, the type of which is to be sought for in the islands of the South Seas." Only a shocking external blow like the potato blight could shake them out of their complacency and alter their ways. The Irish people must learn "to live upon a bread and meat diet, like those of the best parts of England and Scotland." One million Irish people died and another million emigrated during the Great Hunger.[61]

The legacy of Malthus also helped to launch a new idea of nature in the second half of the nineteenth century. Charles Darwin (1809–1882) directly challenged the anthropocentrism of the early modern Cornucopian and Enlightened Scarcity. Instead of celebrating the power of human ingenuity, Darwin declared that natural selection was "immeasurably superior to man's feeble efforts." He rejected the religious anthropocentrism of Genesis and replaced the story of Adam and Eve with a new narrative of material evolution, which demoted humanity to a niche in the vast web of life.[62]

According to Darwin, Malthusian pressures propelled the process of natural selection. Everywhere, animals and plants multiplied without restraint. While humans could curb sexual desire through the faculty of foresight to exercise a preventive check and delay reproduction, other species had no such power. But everywhere in nature, competition over the limited food supply set limits to population. Malthusian Scarcity ruled the natural world. In the constant struggle for life, minute differences provided decisive advantages over competitors. Such inherited modifications gradually gave rise to new varieties and species. Humans could imitate this process only clumsily, through methodical breeding which seized on external traits useful to humans without touching the internal mechanisms that natural selection alone could modify.

One of Darwin's most vivid examples of the superiority of natural processes was the work of earthworms in preparing the land for vegetation. By passing the raw materials in the soil through their alimentary tract, earthworms produced humus at the rate of two inches per year. More or less ignored by human observers, the silent labor of earthworms made possible the entire edifice of civilization. They had "plowed" the land long before the invention of agriculture.[63] Where the physiocrats had seen a partnership between humans and natural processes, Darwin underscored the superior complexity and intricacy of the natural world. Human handiwork was shoddy by comparison. "Nature's productions," he observed, were "infinitely better adapted to the most complex conditions of life, and . . . plainly bear the stamp of far higher workmanship."[64] In the last lines of *On the Origin of Species* (1859), Darwin celebrated the cosmic forces of self-organization in nature:

> Thus, from the war of nature, from famine and death, the most exalted object which we are capable of conceiving, namely the production of the higher animals, directly follows. There is grandeur in this view of life . . . whilst this planet has gone cycling on according to the fixed law of gravity, from so simple a beginning endless forms most beautiful and most wonderful have been, and are being, evolved.[65]

Unlike his contemporaries John Ruskin and Karl Marx, Darwin expressed no anxiety about the destructive consequences of human activities on the natural world. Perhaps it simply did not occur to him that humans might damage the Tree of Life.

The rich legacy of Malthus and Darwin opened up many new lines of inquiry in the late nineteenth century, from ecology and eugenics to anarchist social theory. T. H. Huxley argued that all biological organisms, including animals and humans, were defined by a constant struggle for existence, in which everyone strove to gain as large a share of society's resources as possible. In contrast, the Russian anarchist Peter Kropotkin (1842–1921) dismissed Social Darwinism in favor of a theory of mutual aid.[66] Drawing on evidence from the fields of biology, anthropology, history, and sociology, as well as his extensive travels throughout Siberia, Kropotkin decentered interspecies struggles and emphasized how animals and humans are constantly engaged in various forms of mutual aid. Whether among microbes, animals, or humans, altruism and cooperation drove evolutionary change. Thus, biological life was not simply a never-ending struggle over scarce resources but also defined by a rich tapestry of cooperative practices.[67] Among humans, the principle of mutual aid had been marginalized by the formation of centralized states and the ideology of individualism. Against Malthus and Huxley, Kropotkin argued that competition among people was a highly contingent social dynamic that could be eliminated by the proper organization of society.[68] The challenge facing humanity was therefore to develop new kinds of institutions that would enable people to embrace their cooperative spirit fully, and restore mutual aid as the fundamental social dynamic.

Conclusion

Half a century before the Irish famine, Malthus had predicted that England would face a subsistence crisis within a generation if exponential population growth continued. Yet, contrary to his original prognosis, the disaster occurred in Ireland rather than England—despite the fact that the British population (England, Wales, and Scotland combined) had doubled from 10.5 to 21 million between 1801 and 1851. Just two years after the Irish famine ended, Prince Albert and his industrial and scientific allies organized a celebration of British manufacturing supremacy at the Great Exhibition of 1851. While British industry took pride of place, the space also included exhibits by many European nations as well as the United States, Brazil, and China. More than six million people—a third of the population of Britain—visited the Exhibition during six months between May and October. The event marked the apogee of Victorian industrial might. We might think of the enclosed perimeter of the glass palace as a cornucopian rejoinder to the island anxieties of Malthus. By walking through the halls of the building, a visitor could in the space of a few hours survey the entire scope of the manufacturing economy in all its immense productivity and ingenuity. Political economists had long described the nature of national prosperity through abstract concepts and numbers. The Great Exhibition rendered that intangible concept of abundance into a new form of concrete, personal experience, accessible not just to the middle class but anyone who could afford the one-shilling price of admission. While Malthusian pessimism would persist into the twentieth century, from now on it had to reckon with a novel sense of confidence about the revolutionary consequences of industrialization. For Marx and his followers, technology in the service of socialist revolution would deliver universal abundance and equality. For the marginalist economists, technology in the service of markets promised endless growth and infinite wants.[69]

In the end, neither Ricardo nor Malthus won the argument over Britain's economic development. Contrary to Malthus's forecast, the population of Britain expanded without disaster. Instead, the great famines of Victorian Britain occurred on the colonial periphery in Ireland and South Asia. Ecological and climatological conjunctures like the potato blight and the failure of the monsoon triggered mass starvation in colonies weakened by long-standing exploitation and undemocratic administration. While liberal political economy made strong inroads in policy and politics over the nineteenth century, it was deeply embedded in the power relations of the Empire. Urban consumers were fed and clothed in great part thanks to the British colonies and the wider sphere of informal empire. Even as the famine ravaged Ireland, the island exported wheat and oats for British consumers. Crucial resources also flowed into Britain from the plantation economy of the American South and the wheat belt of the American Midwest. By the end of the century, Britain had become deeply dependent on imports from abroad: 75 percent of all food consumed in the nation was grown outside it. The outsourcing of the food supply also led to ecological consequences. In the early 1860s, annual meat imports stood at five pounds per capita, whereas in 1906 through 1910, this had risen to almost forty-four pounds. An entire global infrastructure of cattle ranches, industrial killing, and refrigerated transport underpinned the British appetite for beef and mutton. This pattern of extensive land use to feed people in the industrial metropoles reshaped the world's ecosystems from Australia to the Pampas.[70]

The British path of development rested on a combination of military might and cheap fossil-fuel energy. By the second half of the nineteenth century, both the Royal Navy and Britain's merchant marine relied on steam rather than sails. This dependence on fossil fuel set a pattern of energy-intensive growth for future powers like the United States and China. The triumph of free-trade imperialism also ushered in the first hint of the shift to come in the earth system. By the third quarter of the nineteenth century—the heyday of Pax Britannica—atmospheric carbon dioxide levels began to diverge from the Holocene pattern of natural variability.

The conquest of Malthusian limits would leave a deep imprint on the planet. In the twentieth century, these disturbances multiplied and accelerated to a point where they began to overwhelm the earth system.[71] Old fears of overpopulation and harvest failure, the hallmark of Malthusian Scarcity, now mingled with novel concerns about the depletion of biodiversity and the inadequacy of carbon sinks—the phenomenon we will explore in Chapter 8 under the name Planetary Scarcity.

6

SOCIALIST SCARCITY

When peace returned to Europe after the Napoleonic Wars, the ruling elite of the great powers convened in the mineral spa town of Aix-la-Chapelle (now Aachen, Germany) to forge the future of the continent. The treaty of 1818 helped lay the groundwork for a reactionary alliance—the Concert of Europe—which was engineered by the Austrian chancellor Klemens von Metternich to check French power, prevent wars in Europe, and suppress political radicalism wherever it might surface. The new system did keep the peace between the nations as intended, but it failed wholly in the task of suppressing dissent. Indeed, one particularly subversive thinker, the British manufacturer Robert Owen (1771–1858), actually used the conference in Aix-la-Chapelle to solicit support for his ambitious reform agenda. He was one of the first figures in Europe to recognize the revolutionary potential of industrial technology. The new manufacturing system could produce misery on an unprecedented scale, he warned, but it might also usher in an age of universal affluence if the relations of production and the property system were transformed. Although Owen's radical proposals were swiftly rejected in the halls of power, his mission provided a new political agenda for critics of the social order. Together with kindred spirits like Charles Fourier (1772–1837) and Henri Saint-Simon (1760–1825), Owen set in motion a wave of reform agitation and communal experiments across Europe. In the hands of Karl Marx (1818–1883) and Friedrich Engels (1820–1895), this critique of industrial society would mature into a sophisticated social theory of capitalism and a worldwide revolutionary movement by the second half of the century.[1]

Owen, Fourier, and Marx, in their own unique ways, put forth a critique of the experience of scarcity that modern capitalism had ushered in. They articulated a new version of the problem that More and Winstanley had recognized—namely, that in capitalism the rich constantly wanted more because their thirst for accumulation was insatiable, and the poor constantly needed more because they owned nothing, or next to nothing. While More and Winstanley had theorized the dynamics of accumulation in agriculture, the nineteenth-century socialists confronted a novel form of exploitation centered on the factory system. According to Marx, who developed the sharpest formulation of this problem, the rich were motivated not by a desire for wealth or consumption, but by a pervasive desire for power. Although the socialists agreed that this condition was intrinsic to capitalism, they also insisted that there was hope for a better future. Human advancement in science and technology had shown what kind of affluence was possible. The socialists believed that the challenge ahead was to transform property relations so that it was no longer true that the poor majority owned nothing and the rich minority kept on pursuing accumulation incessantly. Once genuine human needs were placed at the center of economic considerations and technology was mobilized in the service of all of humanity, what we call *Capitalist Scarcity* would be overcome. In its place, *Socialist Scarcity*—a relation between authentic human needs and an ever expanding material affluence (a hybrid between Finitarian and Cornucopian Scarcity)—would emerge.

Industrializing Europe

Socialist conceptions of scarcity reflected, if only through a glass darkly, the hopes and horrors of Britain's Industrial Revolution. While the term *revolution* implies an analogy with political events like the French Revolution, the transformation wrought by industrialization had deeper, more long-lasting implications for humanity and the natural world than any political metaphor might suggest. We can measure the gravity of these changes through a range of economic and environmental metrics: productivity and real wages; urbanization, population and public health; technological change and energy consumption; novel forms of land use, ecological disturbance, and pollution on a widening scale from the local to the planetary level.

Industrialization opened new worlds of freedom and prosperity for some, while condemning others to stunted lives of degradation and despair. Over the course of the nineteenth century, industrialization made possible a sustained rise in real wages per capita throughout most of Europe and North America. Agricultural productivity also rose steadily thanks to advances in crop use, plant breeding, mechanization, and, by the start of the twentieth century, the invention of synthetic fertilizers. Workers shifted from the countryside and the agricultural sector to toil in manufacturing and the service industry in urban centers. The social cost of economic progress was steep. Mechanization may have raised labor productivity but it destroyed traditional livelihoods in the process. Low wages, long hours, and unsafe conditions made factory labor a constant struggle. The workforce was housed in dismal urban slums plagued by disease and inadequate amenities. When workers resisted such exploitation, their struggle for recognition was met with brutal repression by the owners of capital and the government.

Of all the causes of Britain's industrial revolution, cheap and abundant coal provided the essential driver of economic and environmental transformation. By switching from wood fuel to coal, British society escaped the constraints imposed by exclusive reliance on photosynthetic processes. The high energy density of coal made it a superior source of thermal and kinetic energy for manufacturing processes and resource extraction. Coal was employed to propel steam engines, fire bricks, and smelt iron. By the second quarter of the nineteenth century, coal also fueled a new infrastructure of steam transport on land and sea. Fast and cheap bulk transport paved the way for the emergence of a truly global economy, making far-flung outposts in Argentina and New Zeeland essential suppliers for European household consumption.[2]

The roots of this new energy regime lay in the complex interplay of numerous developments, some technological, others social and political. Long before the advent of the factory system, London had become a fossil-burning city. Sulfur dioxide from the burning of coal turned the sky hazy and caused both respiratory problems and heart disease.[3] Mounting metropolitan demand put pressure on collieries to expand production. Newcomen's steam engine, invented in 1712, found its first economic use as a drainage technology that dramatically expanded access to new coal reserves below the water table. Around the same time, innovations in smelting technology made it possible to produce high-quality iron through the use of coke. Cheap iron production in turn lowered the cost of iron tools and heavy machinery, including steam engines. In the second half of the eighteenth century, water-powered technology revolutionized the output of the textile industry. Soon after, engineers improved the steam engine to increase its fuel efficiency and introduce rotary motion so that it could propel manufacturing processes like cotton spinning.

Such technical milestones in turn combined with favorable social and political forces. The British imperial state and its powerful economic interest groups shaped trade flows and manufacturing opportunities. Imports of Asian semi-luxuries and addictive New World crops stimulated new consumer demand in the middling sorts. When British entrepreneurs set out to imitate Bengal calicoes, their infant manufactures received economic protection from parliament. The British iron industry also benefited from favorable legislation. Meanwhile, imperialism and colonial trade forged captive markets for British manufacturing goods. Another crucial impetus for industrialization came from provincial interest groups. Local landowners and merchants pooled resources to build the British canal network after 1760. These artificial waterways connected coalfields and iron mines with emerging cities, fixing the geography of urban development and the location of heavy industry. By the time cotton textiles emerged as the leading industry of the nation in the 1830s, fossil fuel was already deeply entrenched in the geography and political economy of the nation.[4]

The factory system exploited not just workers but also the environment. Air and water pollution blighted cities like Manchester and Sheffield. Industrialization also set in motion a more profound change that would make itself felt only in the next century. Greenhouse gas from the coal economy gradually began to interfere with the carbon cycle that regulates the climate of the planet. By the last quarter of the nineteenth century, anthropogenic emissions caused the concentration of carbon dioxide in the atmosphere to climb above the pattern of relative stability that had marked the last eleven thousand years. Just at the moment when industrialism seemed poised to conquer the earth, the relative stability of the Holocene epoch came to an end.[5]

Robert Owen and the other socialist writers were keen observers of this new British energy regime even if they did not grasp its long-term consequences for the planet. What struck them above all was the human toll exacted by the industrial economy. Runaway urban growth put pressure on basic amenities such as housing and sanitation. The population of Manchester, the chief industrial city of Britain, doubled in the first twenty years of the nineteenth century and again between the 1820s and 1850, growing to 400,000 people. Unregulated industry and negligent public provisions produced nightmarish conditions for the urban working class from overworking, inadequate diet, overcrowding, rampant disease, and a paltry water supply. Friedrich Engels described the pinched existence of the lower working class—dressed in cheap cotton and subsisting on a diet of "bread, cheese, porridge, and potatoes" with "weak tea."[6] Biophysical statistics tell a grim tale of systematic exploitation and neglect. Life expectancy averaged 29–30 years in towns above 100,000 between 1830 and the 1840s. In 1841, average life expectancy stood at 25.7 years in Liverpool and 25.3 in Manchester. Tuberculosis, cholera, and scarlet fever ravaged the population of Manchester. "If one roams the streets a little in the early morning, when the multitudes are on their way to work" Engels reported, "one is amazed at the number of persons who look wholly or half-consumptive"—throngs of "pale, lank, narrow-chested, hollow-eyed ghosts."[7] The social impact of the early stages of industrialization was nothing less than catastrophic. By the 1830s, evidence of urban crisis began to enter into public consciousness through newspaper reports and official inquiries. Eventually, the British legislature followed suit with limited reforms, curtailing child labor and reducing the length of the working day. Medical inquiries and increased public spending slowly improved living conditions. But for much of the nineteenth century, the prospect for the working class looked increasingly dire. Fears of catastrophic deterioration set the stage for a variety of socialist critiques of capitalism.

A British Utopian Socialist—Robert Owen

Robert Owen's project of universal reform began on the factory floor. After making a name for himself in the cotton business in Manchester, Owen became the manager of his father-in-law's manufacture outside Glasgow, known as New Lanark. Here he faced the pressing challenge of recruiting labor for the unfamiliar and often dangerous tasks of mechanical spinning. His father-in-law, David Dale, had resorted to recruiting migrants from the impoverished Highlands and then from the orphanages in Glasgow. But these novices proved difficult to manage. To improve work habits, Owen began to experiment with new kinds of education and welfare for the workers. Over time, the lessons from these trials matured into a full-fledged philosophy of education, which Owen published in 1813 under the title *A New View of Society*. In contrast with his contemporary Malthus, Owen argued that human nature was radically malleable. Through enlightened pedagogy, working-class children could be taught to transcend their social condition. Factories thus served both as laboratories for a new kind of humanity and as the source of unprecedented material prosperity.

On his way to the Concert of Europe in 1818, Owen stopped in the city of Frankfurt to meet with diplomats and government officials. Here he wrote and printed two remarkable memorials addressed to the governments of Europe and America in the name of Frankfurt's working classes. Published simultaneously in English, German, and French, the texts doubled as political manifestos and prescient descriptions of the new British manufacturing economy. Owen's own experience in the cotton industry inspired his understanding of technological and social change. He singled out the inventions of James Arkwright's spinning machinery and James Watt's improved steam engine as examples of what he called a new form of "scientific power."[8] This technical advance significantly increased the productive power of the British industrial worker. While the size of Britain's labor force had grown from 3.75 million in 1792 to 6 million in 1817, the most astonishing increase in the nation's productivity came from new machines fueled by waterpower and coal. By 1792, steam engines, water looms, and other machinery had increased the productive power of the workers threefold, equivalent to what a labor force of 11,250,000 men could produce. Twenty-five years later, the multiplication of "scientific power" had increased productivity of labor by a factor of thirty (which Owen suspected might in fact be an understatement). This meant that Britain was now producing on a level that with the old technology would have required 200 million workers. Such a stunning advance was for Owen only the beginning of a process that might actually eradicate, once and for all, food and housing insecurity, and thereby change the meaning and experience of scarcity. When he republished the 1818 memorials at the end of his life, he added the following note to the original text:

> N.B.—Now, in 1857, the scientific, mechanical, and chemical productive powers of Great Britain may be estimated as equal to at least *one thousand millions* of men; or, in proportion to the present population, as *thirty three to one;* and *this is continually increasing, and is capable of illimitable augmentation.* R.O.[9]

Although the term "Industrial Revolution" would not enter into the English language until a generation later, Owen's 1818 pamphlets identified several of the features that we now take as essential traits of industrialization: new machines, new science, new ways of organizing labor, new markets, and most importantly, new energy sources. According to Owen, the quantum leap in productive capacity also extended into agriculture: "man knows not the limit to his power of creating food."[10] Chemical expertise promised novel ways of augmenting the food supply. Owen's sanguine expectations resembled to a certain degree the Hartlibians' technological millenarianism from the seventeenth century. Against forecasts of diminishing returns, Owen insisted that "the population of the world may be allowed naturally to increase for many thousands of years."[11] In what could only be seen as a snub to Malthus, Owen predicted that productive capacity would in fact outrun population growth: Britain "could soon, by the help of science, supply the wants of another world equally populous with the earth."[12] Society had entered a stage of history marked by continuous technological change and economic growth. The future was dynamic and open-ended.

Humanity had arrived at a new dawn when "scientific power" would bring about universal affluence. "Riches may be created in such abundance and so advantageously for all, that the wants and desire of every human being may be over-satisfied."[13] The new system of manufacturing offered an opportunity to alleviate misery at home as well as abroad. Britain ought to "extend the knowledge which she has herself acquired of creating wealth or new productive power, to the rest of Europe, to Asia, Africa, and America."[14] To that end, a new principle of political economy was necessary, one that would regulate the distribution of productive power: "It is the grand interest of society to adopt practical measures by which the largest amount of useful and valuable productions may be obtained at the least expense of manual labor and with the most comfort to the producers."[15] In this soaring vision, the whole world would soon enter into an age of universal abundance. Prosperity combined with a more educated and enlightened population would usher in an era of Socialist Scarcity.

However prescient and laudable, such hopes quickly proved premature when Owen's 1818 mission ended in abject failure. Later radicals thought the scheme was flawed from the outset by Owen's unwavering commitment to peace and social harmony. Despite his sharp criticism of the manufacturing system, Owen still believed in incremental improvement, thus obviating the need for a revolution. The purpose of the 1818 memorials was to win over governments and elites through rational persuasion and top-down reform. It is true that Owen became more open over time to the value of large-scale grassroots organizations like the Chartist movement, but he never abandoned his distaste for insurrection. A generation later, Marx and Engels dismissed Owen as a utopian thinker who failed to grasp the true logic of capitalism and the need to overthrow the bourgeoisie. Yet, for all the criticism he received, Owen's role in laying the groundwork for socialist thought should not be underestimated.

A French Utopian Socialist—Charles Fourier

While Owen recast the notion of scarcity by concentrating on improving the conditions of factory work and using the new technology to greatly expand the wealth available to properly feed, clothe, and house the entire population, his fellow utopian socialist Charles Fourier, one year younger, focused on rethinking human desires, emphasizing the richness and diversity of human passions. The Lyon resident was deeply familiar with Owen's writings and his practical experiments at New Lanark. In the spring of 1824, Fourier wrote to Owen, asking for permission to join a new venture he was planning in Motherwell, Scotland. Owen responded politely, but stopped short of offering him a position. A few months later, Fourier once again wrote to Owen, this time via Owen's French-speaking disciple Philip Skene. Fourier offered assurance that, if he were invited to join Owen's project, success would follow: "My intervention will be the guarantee for a brilliant future for you and even more for Monsieur Owen, who as founder of the association will win the title of Social Messiah."[16]

Though born into a merchant family, Fourier swore "an eternal hatred of commerce" from an early age.[17] While money troubles frequently forced him to return to the mercantile professions and take work as a cashier, bookkeeper, shipping clerk, or commercial broker, this did not stop him from developing an acerbic critique of modern commercial society. Fourier enumerated 144 permanent failings of commercial "civilizations."[18] Drawing inspiration from Jean-Jacques Rousseau, Fourier argued that humanity had organized life within a set of social institutions that had subsequently turned on people and made their lives intolerable. Religious authorities and Enlightenment philosophers alike had completely misunderstood the human condition and therefore forged ideas and ideologies that legitimized an economy of waste and a culture of human subjugation. To emancipate humanity from this yoke, so that a new era of "Harmony" could eventually be ushered in, an entirely new social science was required—one that was grounded in an honest assessment of human passions. While earlier thinkers, such as Bernard Mandeville, had proclaimed that philosophers had yet to study people the way they really were and not the way they ought to be, Fourier insisted that they, too, had fallen short by focusing so tightly on people's pursuit of pleasure through consumption. Fourier argued that, in the same sense that Isaac Newton had created a new science on the basis of his discovery of gravity, he himself was on the cusp of forming a social science based on a brand-new understanding of the passions. He called his theory the "Geometric Calculus of Passionate Attraction."[19]

Fourier divided the twelve passions into three categories: sensory passions, affective passions, and distributive passions. *Sensory passions* captured the five pleasures humans derive from *sight, smell, touch, hearing,* and *taste.* These passions were directed toward the enjoyment of objects. The four *affective passions,* in turn, were directed toward other people and the pleasures derive from *friendship, love, ambition,* and *parenthood.* These passions tended to be connected to the human lifecycle,

with children drawn to friendship, young men and women inclined toward
love, mature people embracing ambition, and the older generation focusing on family. In commercial "civilization," Fourier lamented, mature
people had managed to impose their passion on the rest of society, contributing to the repression of the other affective passions. The final three passions that animated humans, which fell under the rubric of *distributive passions,* were the *cabalist, butterfly,* and *composite passions.* The cabalist
passion referred to the penchant for conspiracy, calculation, and competition. While these often sparked antisocial behavior in "civilization," they
might, in a differently organized society, inspire people to work hard and
pursue perfection. The butterfly passion referred to people's desire for a
variety and diversity of pleasures. This passion was particularly suppressed in marriage and in work, Fourier's two main bugbears, about
which we will say more below. Also severely restrained in civilization, the
composite passion was a mixture of pleasures of the senses and pleasures
of the soul. The love of food, for example, becomes a refined gastronomical
pleasure when enjoyed in the presence of good friends, all sharing a soul-
ful interest in culinary delights. Lovemaking that fused physical and spiritual connections also represented a composite passion.

These twelve passions constituted the source of all our happiness
and, as such, there was nothing intrinsically wrong about pursuing any
of them, according to Fourier. The reason that philosophers had demonized
them in the past and sought to impose limitations was that they often
turned destructive when corrupted by poorly designed institutions, such
as private property, profits, and monotonous work. The key was to find a
social organization that was in harmony with people's psychology—one
that liberated the passions so that people could live free, happy, and creative lives.[20] Liberation of the passions, not self-denial, was the point of
Fourier's ideal society. By recognizing that pleasure from consumption
constituted only a small subset of the varied pleasures humans are capable of enjoying, Fourier argued that it made little sense to theorize the
human condition as one characterized by material scarcity. So many of
the objects of human passions had absolutely no material limitations.

Fourier proposed that humanity should be sorted into small, largely
self-sufficient societies on the basis of their unique composition of the
twelve passions. He estimated there were 810 distinct personality profiles.
Fourier suggested that people should be combined into harmonic groups—

voluntary associations of people from different classes and ages who shared complementary traits. These groups, in turn, would be combined into so-called phalanxes of 1,620 people, with two of each profile. Once these societies were formed, people could pursue their individual mix of pleasures in a manner that fostered both their own happiness and that of their society. Political authority would become superfluous, as the underlying causes for antagonism would be eliminated. Only a nominal governance structure was required, with a regency or council making day-to-day decisions regarding scheduling and sequencing of work.

The phalanxes were to be located in rural settings, in relatively hilly landscapes, with plentiful water supply and fertile land. Each community would have an elegant central building, containing workshops, entertainment rooms, playrooms, dining rooms, libraries, and kitchens. The governing logic of the architectural design was to form communities and to eradicate waste. For example, by having communal kitchens and dining rooms, the preparation and consumption of food would free women from toiling endless hours in isolation and save families from having to endure the daily mundane dinner. Instead, women would be free to join the labor force and the meal would turn into a sensuous and social experience. The art of cooking would become the most revered skill and each member would have daily access to "twelve types of soup, twelve varieties of bread and wine, and twelve dressings for meat and vegetables."[21]

The setup of the phalanxes was designed to emancipate humanity from two of civilization's most repressive institutions: work and marriage. Work, Fourier argued, ought to be organized collectively, to reduce waste and to promote collaboration. No longer would there be scores of family farms, each with its own plot, tools, and fertilizer. Instead, all the people in the community would be engaged in the production of food, shelter, comforts, and amusements, using collectively owned land, tools, and equipment. The parasitical part of the population, who lived off other people's work, estimated by Fourier to be as high as two-thirds in "civilized" societies, would no longer squander their productive energies, but would join the communal endeavor. This included three-quarters of children and servants, all women formerly engaged in household labor, the idle rich and idle poor, and most soldiers, merchants, and government functionaries. While everyone would be engaged in productive pursuits, no one would feel subjugated or exploited. Instead of a form of

H. Fugère, *Vue d'un phalanstère, village français [composé] d'apres la théorie
sociétaire de Ch. Fourier,* nineteenth century, date unknown. The spatial
organization of social life in Charles Fourier's Phalanxes was critical to the revolution
in human passions. Everything was neatly planned for the full flourishing of human
faculties. *Credit:* 860.05, Houghton Library, Harvard University.

imprisonment, work was now one of the main sources of pleasure and
personal fulfillment.

All work would be conducted in units chosen voluntarily on the
basis of friendship and attraction. Tasks were varied frequently, ideally
around eight times per day. This meant that on any given day a person
would work for an hour or two in each of several settings: the stables, the
gardens, the vegetable fields, the forest, the barnyard, the irrigation system, and the factory. Most of the workday was taken up by agricultural
work, which Fourier deemed much more satisfying and pleasant than
manufacturing.[22] Also contributing to the workers' appreciation of work
was the fact that wages would be abolished. Instead, the phalanx would operate as a joint stock company, distributing its profits according to the
following rule: five-twelfths to labor, four-twelfths to capital, and three-
twelfths to talent. This remuneration scheme, combined with the promise that each member of the phalanx was guaranteed a basic minimum of

food, clothing, and subsistence, eliminated the primary source of in-equality and eradicated, once and for all, abject poverty. In Fourier's mind, poverty was the most egregious stain on "civilization." Fourier also contemplated ways to rid the mercantile world of dishonesty, which would go a long way toward eliminating fraud and corruption, the most common causes of high prices and shortages. Together with the flourishing of humanity's passion, this rearrangement of the economy would emanci-pate humanity from the experience of scarcity, whether the version expli-cated by Hume, Malthus, or later by Alfred Marshall and the Marginalists.

The second "civilized" institution that would be excluded from the phalanxes was marriage. Fourier saw marriage as a social sickness in that it imposed stifling strictures on half the world's population. When women married they became the property of their husbands and were over-whelmed by the tasks of housekeeping, cooking, and child-rearing, losing their freedom to pursue their own unique sets of passions. Together with their husbands, they entered a world in which it was prohibited to embrace erotic love freely. As the vast majority of people's pleasure profiles included lovemaking and only a few people exhibited monogamous tendencies, as witnessed by the frequency of infidelities and cuckoldry, the institution of marriage transformed one of the richest sources of pleasure, the compos-ite passion, into boredom and monotony.[23] By turning lovemaking into an obligation or even servitude, marriage erased the mystical and alluring features of the erotic pleasure, and therefore drove both men and women to pursue illicit liaisons. The butterfly passion was just too strong to be subdued. Fourier believed that a diverse set of erotic gratifications were es-sential to both personal happiness and social harmony. Indeed, lovemak-ing should be a central everyday activity and not be hidden away in the private quarters of married couples. Since everyone had their own per-sonal sexual proclivities, Fourier suggested that a card file be kept to sim-plify the matching process and that visiting travelers wear an insignia symbolizing their preferences—much in the spirit of modern-day dating services. Finally, in the same sense that Fourier prescribed that each member should be entitled to a minimum level of material comforts, so too did he believe that every person was entitled to a minimum of erotic pleasure. The abundance of erotic passion was an important ingredient of the transcendence of Capitalist Scarcity. Once people pursued a mix of all twelve passions, the focus on material wealth would fade.

Fourier's predominantly agrarian phalanxes shared much common ground with More's Utopia and the vision of bucolic bliss promoted by the Romantic movement. Although he was not entirely opposed to modern industrial technology, he wanted to eliminate factory labor as a person's sole occupation; people ought to spend the bulk of their lives outside, immersed in the natural world, enjoying its beauties. Fourier had been fascinated by flowers since his boyhood in Lyon. Young Charles would fill his room with flowers, initially planted in pots, but later in flowerbeds on the floor. His sister reminisced, "On each side the floor was adorned with the prettiest flowers, tuberoses, tulips, and others. Of course, when he left us and the soil was removed from the top of the floorboards, they had all rotted and it was necessary to do the whole room over again."[24] In a moment of mystical reflection about the interconnections between nature, society, and the celestial sphere, Fourier speculated that the future social organization would have salubrious effects on the environment and the climate. Once the noxious gases or aromas emitted by the industrial economy came to an end, the sun would once again be able to heat the earth properly and increase its fertility. This benign form of global warming would have all kinds of wonderful effects, enabling farmers to cultivate grapes in St. Petersburg and oranges in Warsaw.[25]

Contrary to most contemporary philosophers, Fourier did not see the necessity of stifling any of the human passions, either through self-imposed restraints or through externally imposed laws, secular or religious. No one should be forced to "live on 'republican cabbage,' 'Spartan gruel,' and self-restraint."[26] If people wanted to pursue unlimited pleasures of the senses, it should be their right to do so. While there were indeed some people who were inclined to chase satisfaction through the consumption of luxuries, most people had a more balanced set of desires; in addition to consumption they also derived enjoyment from creative work, sumptuous meals, lovemaking, and friendships. Life in the phalanx was not a quest for never-ending consumption or endless accumulation, as in Mandeville's vision of society. Happiness was not a simple one-dimensional pursuit of consumption. Moreover, given that all members were engaged in productive pursuits and each member of the phalanx was guaranteed a minimum amount of material comforts, the community would eradicate the experience of Capitalist Scarcity and usher in an era of Socialist Scarcity.

Marx's Critique of Capitalist Scarcity

When revolutions broke out across Europe in 1848, the brilliant young student Karl Marx found himself swept away by the promise of political and social transformation. He decided that he would try to use philosophy not just to understand the world but to change it.[27] Karl Marx was born into a family of rabbis in Trier in 1818. His father, Heinrich, converted to Christianity once he returned from his legal studies in France, realizing that being Jewish would likely harm his career possibilities in Prussia. Once Karl enrolled at university, the expectation was that he would follow in his father's footsteps and study law, but he kept getting sidetracked. The first intellectual pursuit that knocked him off-course was poetry—in particular, Romanticism. Next, through his encounters with the Young Hegelians in Berlin, he took a deep interest in German idealism. After completing his doctorate in 1841, Marx began to publish articles advocating for classical liberal ideals, such as freedom of the press and freedom of expression. It was only once he arrived in Paris that his radical political journey began in earnest. As he read French authors like Fourier, Saint Simon, and Pierre-Joseph Proudhon, and paid close attention to ongoing working-class revolts throughout Europe, Marx began to develop his own critique of capitalism. It bears emphasizing that Marx's oeuvre was almost entirely dedicated to the criticism of capitalism as it was, and included only sporadic glimpses of how a post-capitalist society might look.

The more Marx learned about the labor conditions in mines and factories, the poverty and death in the factory towns, and the striking disparity in the distribution of wealth, the more he became convinced of the necessity for radical change. He was particularly incensed by the manner in which capitalism had turned work into labor; a genuinely human expression of creativity had morphed into a brutal form of social control. For Marx, labor was the fundamental means whereby the small minority of capitalists maintained their domination over the bulk of the population. By creating a system of private property and making sure that all essential features of society were privately owned, the ruling elite forced the propertyless to sell their labor power to gain access to food and shelter. The working classes were thus forced to dedicate the better part of their days, weeks, months, and years to wage-labor to eke out even a meager living. During the time they spent at work, as well as the time spent preparing and recovering from work, the laboring men and women were under the physical and psychological dominion of their employers. In their joint *Communist Manifesto,* Marx and Engels described "the miserable character of this appropriation, under which the labourer lives merely to increase capital, and is allowed to live only in so far as the interest of the ruling class requires it."[28]

In the early 1840s, Marx had begun to develop his theory that labor constituted the primary form of social control in capitalism, jotting down his thoughts in a set of notebooks that were posthumously published as the *Economic and Philosophical Manuscripts of 1844.* He noted that because the worker's "own activity is to him related as an unfree activity, then he is related to it as an activity performed in the service, under the dominion, the coercion, and the yoke of another man, . . . someone who is alien, hostile, powerful, and independent of him."[29] This other man, the capitalist, was now "elevated to the position of a commanding lord." As for the capitalist, "through his wealth, he has gained power," Marx continued, and in putting this power to use, he "effectively bends to his purpose those he employs."[30] As a class, the capitalists thus asserted their dominance over the working class through the imposition of labor.

To ensure that labor constituted an active transference of power, capitalists arranged things so that workers were unable to exercise control over the labor process, the products they produced, their relationship to their fellow workers, or their own creative potential (which Marx referred to as their "species-being"). Marx described the workers' alienation from work, product, fellow workers, and species-being as both a process of dehumanization and disempowerment. While laboring in a factory under terrible conditions, for fourteen hours per day, undoubtedly damaged a worker's inner life—psychologically, emotionally, and intellectually—such conditions kept capitalists in control of the workers. Under capitalism, therefore, labor was the activity by which workers were turned into a class; this was an active process of appropriation and subjugation. As such, the imposition of alienating labor was not, first and foremost, a mean-spirited plot on the part of the capitalists to make life miserable for the workers—although it did that, too—but rather a strategy designed to reproduce their own social power.

Marx did not use the term alienation with any frequency after the 1840s, but he continued to train his analytical focus on the imposition of

labor as the primary form of social control throughout the rest of his voluminous writings.[31] Once he adopted the labor theory of value from David
Ricardo and the classical economists, he used the term *abstract labor* to re-
fer to alienating labor. The "abstraction" involved in abstract labor, in other
words, was not a theoretical construct but reflected the homogeneity
of labor in capitalism. "This reduction," Marx wrote of abstract labor,
"appears to be an abstraction, but it is an abstraction which is made every
day in the social process of production."[32] Marx's version of the labor theory of value, measured the extent to which capitalists as a class were able
to continue imposing control over the working class and reproducing class
relations.

Viewed from the individual capitalist's point of view, the accumulation process was about the amassment of as much money as possible. To
Marx, money existed as a representation of value and therefore the em-
bodiment of power. It was through money that "social power becomes the
private power of private persons."[33] Moneymaking was a limitless process,
"for the valorization of value takes place only within this constantly re-
newed movement."[34] To reproduce their power, the capitalists had to
constantly reinvest their money. If they spent too much of it on frivolous
consumption, they soon lost their capacity to reproduce their power. Only
unsuccessful capitalists succumbed to consumerism. The successful capitalist was well aware that the only way to retain his power, as an individual and as a member of the capitalist class, was "by means of throwing
his money again and again into circulation."[35]

Marx's version of Capitalist Scarcity was fundamentally different
from the other conceptions that we have discussed in this book. He did
not, first and foremost, care about the quantity of goods available, the
desires that shaped consumer behavior, or the depletion of natural resources. He focused on the sphere of production not because it was the
site of production of useful material goods, but rather because it was the
space in which control was imposed through alienating labor.[36] The lim-
itlessness that characterized capitalism and served as the source of scarcity was the endless quest for power on the part of the capitalists. Since
power was never absolute, always relative, there was never an end to its
pursuit. Capitalism was a never-ending circuit of money, labor, and power.
To Marx, if this circuit came to an end, it would spell the end of capitalism itself.

While the capitalists' pursuit of power was boundless in theory, it was always limited in practice by the self-activity and resistance of the working class. This was *the* epic drama at the heart of capitalism, Marx argued. He saw the workers' struggles for higher wages, more free time, and more control over the labor as stratagems in the class war. The more that workers were able to retain control over time, energy, resources, and space, the greater their power they had over their own lives. Marx believed that the working classes were always engaged in a fight for their own autonomy and independence. It did not require instruction or education for workers to resist; subjugated people intrinsically refuse their domination and struggle against their superiors. The more successful the workers were in asserting their independence, the more they were able to participate in the formation of a more convivial society. This so-called self-valorization of the workers had the potential, Marx argued, to break the control of the capitalists and put an end to the conditions whereby a tiny minority maintained its dominion over the vast majority. The tension in Marx's notion of Capitalist Scarcity was between the capitalists' infinite quest for power and the workers' incessant struggles for liberty and freedom.

Locked in a life-or-death struggle with the working class, the capitalists pursued every available strategy to reassert their control over the workers. They tried to lower wages and extend the working day, but the most effective weapon at their disposal, Marx argued, was the introduction of new machinery. Drawing on Engels's recently published book about the factory conditions in Manchester, Marx wrote in an 1846 letter to his friend, Pavel Annenkov, a Russian literary critic and good friend of the anarchist Mikhail Bakunin, that "since 1825, the invention and application of machinery have been merely the result of the war between employers and workers."[37] He exemplified this tendency in *Capital*, suggesting that "spindles, looms and raw material are now transformed from means for the independent existence of the spinners and weavers into means for commanding them."[38] Self-acting mules and power looms were implemented to replace rebellious spinners and troublesome weavers. Indeed, Marx noted, "by means of machinery, chemical processes and other methods, [modern industry] is continually transforming not only the technical basis of production but also the functions of the worker and the social combination of the labour process."[39]

Introducing new machines was partly about replacing workers, but more importantly it was about redesigning the labor process so that the machine could help restore order and authority. "In machinery," Marx argued, "objectified labour confronts living labour within the labour process itself as the power which rules it; a power which, as the appropriation of living labour, is the form of capital."[40] The worker was forced to confront the machine as dead or zombie labor, an alien power. Once again referring to the isomorphism between alienated and abstract labor, Marx wrote, "The worker's activity, reduced to a mere abstraction of activity, is determined and regulated on all sides by the movement of the machinery, and not the opposite. The science which compels the inanimate limbs of the machinery, by their construction, to act purposefully, as an automaton, does not exist in the worker's consciousness, but rather acts upon him through the machine as an alien power, as the power of the machine itself."[41] Machines therefore contributed to the capitalist aim of transforming work into abstract labor, turning the worker into "an appendage of the machine."[42]

Marx's Vision of the Post-Capitalist World

Marx made a conscious decision not to write much about the design of post-capitalist societies. He did not see it as his responsibility to prescribe what the future should look like. During his formative years in Berlin, he wrote in a letter to one of the young Hegelians, Arnold Ruge, that socialists should be engaged in the "ruthless criticism of all that exists" rather than naively "constructing the future and settling everything for all times."[43] But we can still see glimpses here and there of what he hoped the future would bring. Similar to Fourier, he wanted people to develop their talents in a variety of pursuits; it should be "possible for me to do one thing today and another tomorrow, to hunt in the morning, fish in the afternoon, rear cattle in the evening, criticise after dinner, just as I have a mind, without ever becoming hunter, fisherman, herdsman or critic."[44] To reach this goal, it was obviously essential to get rid of alienated labor, but also to abolish private property. Marx, however, did not want to do away with *all* property, only bourgeois property. Communism, he argued, would not deprive anyone from mixing their labor with nature or engaging in genuine collaboration with others; all it would do would be "to deprive him of the power to subjugate the labour of others."[45]

Marx was also convinced that in the absence of capitalist class relations, technology could be turned into a force for good. Much like Owen, he was profoundly impressed by the massive productivity unleashed by the factory system. It had been the first mode of production to show the true power of mankind over the earth. New sources of energy, new materials, and new machines had given humanity a power over nature that went far beyond what Bacon and the Hartlibians had envisioned. As he and Engels wrote in the *Communist Manifesto,* the bourgeois mode of production had accomplished "wonders far surpassing Egyptian pyramids, Roman aqueducts, and Gothic cathedrals."[46] However, only if people managed to emancipate themselves from the capitalist mode of production, would it be possible to put technology to use in the service of human progress.[47] In short, Marx wanted to retain humanity's power over nature, while eliminating the capitalists' power over the workers.

Referencing Aristotle's famous vision of self-acting technology with shuttles weaving fabrics on their own and plectrums playing the harp without human involvement, Marx envisioned a future in which the magnificent power of the machine could be marshalled in support of humanity's betterment. Indeed, there were already signs during Marx's lifetime that the making of machinery was itself becoming mechanized. "Cyclopean machines" were used to produce ever more complex and sophisticated machines, such as railways and ocean steamers.[48] Scientific expertise coordinated the operation of the system, but the actual work process required less and less human input. Marx described the new order with a mix of awe and revulsion: "Here we have, in place of the isolated machine, a mechanical monster whose body fills whole factories, and whose demonic power, at first hidden by the slow and measured motions of its gigantic members, finally bursts forth in the fast and feverish whirl of its countless working organs." This nightmarish world marked the triumph of "production by machinery."[49] But it also opened the possibility of a new social order whose goal was to humanize the "mechanical monster" and make sure that technology was employed in the interest of all of humanity.

Humanity would continue to "wrestle with Nature" to satisfy their evolving wants, but they would now be aided by the forces of production that had been liberated from the service of domination. In a society in which private property and alienated labor had been abolished and technology was employed for the benefit of all people, the experience of Capitalist Scarcity would be transcended. The incessant quest for capital accumulation on the part of the vampiric capitalist would be eliminated once and for all. In its place, people's genuine physical, social, and aesthetic needs, none of which were intrinsically boundless, would guide production. This did not imply that society would become static; Marx recognized that human creativity and imagination would always create new wants. People would "rationally regulat[e] their interchange with Nature, bringing it under their common control," which would enable them to fulfill their wants with "the least expenditure of energy and under conditions most favourable to, and worthy of, their human nature."[50] Freed from the compulsion to produce without limits, workers would now produce to meet concrete use values, which would always leave them with ample leisure. By providing people with an abundance of free time, post-capitalism would bring about true freedom and genuine autonomy, or so Marx believed.

The reorganization of the productive apparatus around genuine human needs would enable a "fully developed humanism," which for Marx meant "a fully developed naturalism."[51] But he did not volunteer much about his views on the relationship between the economy and the environment. To the extent that Marx explored the natural world at all, he did so through the lens of actual human work or "living labor"—not abstract labor. He saw a distinctive role for nature in all production processes. Paraphrasing William Petty, he observed that "labor is the father of material wealth," while "the earth is its mother."[52] Only through human work did nature come alive and gain purpose: "Living labor must seize on these things, awaken them from the dead, change them from merely possible into real and effective-use values."[53] Marx also recognized that the production of use-value was not simply a one-way process of appropriation. What humans extracted from nature in the form of food and clothing and other raw materials must be replenished and returned to the soil. This relation of metabolic exchange (*Stoffwechsel*) was an eternal condition common to all of humanity. But the endless circuit of accumulation set in motion by capitalist production had disturbed the "metabolic interaction between man and the earth." Such a rift stopped the "return to the soil of its constituent elements" and robbed the land of its "long-lasting . . . fertility."[54] The rift between humans and the earth demanded radical action. In a post-capitalist society, there had to be a "conscious technological application of science" to both agriculture and manufactures, thus creating the "material conditions for a new and higher synthesis."[55]

Marx's defense of scientific agriculture went hand in hand with a complete rejection of Malthusian political economy. The "abstract law of population" existed "only for plants and animals and even then only in the absence of any historical intervention by man."[56] Malthus's idea of natural limits to population had no truth in itself. It had become popular, Marx insisted, only because it was socially useful to the British oligarchy.[57] Marx's angry attack on Malthus was also a dismissal of natural theology. He had no patience with talk of divine design. He also resisted the notion that nature had any intrinsic aesthetic meaning or a higher purpose. Despite his early forays into Romanticism, the legacy of romantic conceptions of scarcity from Jean-Jacques Rousseau to John Ruskin, which had inspired both Robert Owen and Charles Fourier, made little impression on Marx. While Marx read widely in the natural sciences, little of this knowledge made it into the pages of *Capital*.[58] There was no room in his political economy for the restoration of rural traditions or the preservation of wild things. Marx was an urban creature through and through, whose experience of the natural world was limited to Sunday picnics on Hampstead Heath with his family. While Marx's contemporaries had already begun to warn about the dangers of deforestation, climate change, chemical pollution, and the extinction of wildlife, Marx's view of nature remained grounded in his Cornucopian faith in the power of science and technology to master the natural world. He believed that technology would enable the conditions for Socialist Scarcity, a world in which humanity would for the first time in history become truly free.

Conclusion

Owen, Fourier, and Marx envisioned a future of Socialist Scarcity, in which human passions were redirected away from accumulation while technology kept on improving the ability of society to meet all genuine human needs. This version of scarcity incorporated both Finitarian and Cornucopian elements. They were all highly critical and at the same time optimistic about the mode of production introduced by capitalism. Their Cornucopian assumption about industrial technology convinced them that the enormous productive power unleashed by the bourgeoisie had the capacity to generate enough material wealth to satisfy all human needs and to free people from drudgery. Side by side with this Cornucopian faith in technology, the socialist thinkers also envisioned the emergence of a new humanity, one that was free from the inauthentic desire for infinite consumption and repressive power. After the revolution, people would have the leisure to cultivate new passions, engage in creative work, and enjoy convivial social relations. To a certain extent, Socialist Scarcity therefore had some important commonalities with Enlightened Scarcity, in the sense that both concepts put a lot of stock in the development of technology and the transformation of the mind. But there were also, of course, sharp differences, not the least of them being the socialists' ambition to abolish the forms of property and labor at the center of Enlightened Scarcity.

While Marx's vision of freedom lingered in some quarters, it was his critique of capitalism that received the most attention and continued to inspire left-wing thinkers around the globe after his death in 1883. Friedrich Engels, who outlived his best friend by twelve years, played a formidable role, together with his German compatriots Karl Kautsky and Eduard Bernstein, in shaping Marxism after Marx. Others, such as the Russian anarchist Peter Kropotkin and the German radical Rosa Luxemburg (1871–1919), reacted to what they perceived to be a misguided direction of the socialist movement. Luxemburg, who saw the working-class movement as a decentralized force—"It flows like a broad billow over the whole kingdom, and now divides into a gigantic network of narrow streams; now it bubbles forth from under the ground like a fresh spring and is now completely lost under the earth"—was discouraged by the power assigned to the party's central leadership.[59] She, together with Marx, would have been even more discouraged by the turn Marxism took after Vladimir Lenin came to power in the USSR. In one of Lenin's first articles, published in *Pravda* in April 1918, he embraced the American system of scientific management known as Taylorism—the efficiency program that would soon give rise to Fordism. Lenin informed his readers that Frederick Taylor's approach of sending armies of industrial engineers with stopwatches and clipboards into factories to study ways to eliminate "superfluous and useless motions" had yielded crucial insights about "the most correct methods of work, the best systems of accounting and control."[60] This led Lenin to conclude, "We must introduce in Russa the study and teaching of the new Taylor system and its systematic trial and adaptation."[61] The aim of the Revolution was clearly not about transcending the capitalist culture of infinite growth and eliminating exploitative forms of labor. "The possibility of socialism," Lenin declared, "will be determined by our success in combining the Soviet rule and the Soviet organization of management with the latest progressive measures of capitalism."[62]

Once Joseph Stalin came to power in 1922, economic growth became the official mantra of Russian state-socialism. Five-year plans, collectivization, and Stakhanovism were all designed to increase productivity in both agriculture and manufacturing, so that the Soviet Union could maximize its economic growth. The quest for economic expansion was as intense in the East as it was in the West. While the state put strict limits on people's consumption patterns, the infinite quest for more economic growth created conditions that best resembled what we have referred to as Capitalist Scarcity. No cost was too great to bear to obtain the highest growth figures, in terms of workers' exploitation and in terms of environmental destruction. In its effort to develop heavy industries between 1950 and 1991, the USSR increased its carbon dioxide emissions from 12 percent to 16 percent of the global total.[63] To keep up with the United States, moreover, the Soviet Union pushed its energy infrastructure to the brink, eventually triggering an environmental and human disaster with the nuclear reactor core fire at Chernobyl in 1986. Socialist Scarcity never had a chance in the Soviet Union.

7

NEOCLASSICAL SCARCITY

The 1851 Crystal Palace Exhibition in London signaled the beginning of a new era of abundance in world history. The massive glass and cast-iron frame erected in Hyde Park for the five-month exhibition displayed some one hundred thousand objects, organized in four sectors: raw materials, machinery, manufactures, and fine arts. The exhibition captured the air of excitement about the world of goods created by the Industrial Revolution. Yet, many British people were not satisfied with simply enjoying more, better, and cheaper goods. As noted in Chapter 6, they mobilized for greater democratic representation and improved labor conditions. Their struggles stoked genuine fears of revolution, leading Karl Marx and Friedrich Engels to declare in 1848 that "a specter is haunting Europe—the specter of communism."[1] The tension between increased consumer affluence and the intensification of political agitation was clearly felt at the Crystal Palace Exhibition. Organized by Prince Albert and the Royal Society for the Encouragement of Arts, Manufactures, and Commerce, the Exhibition was an exuberant celebration of modern technology and consumerism, and yet some commentators worried that the gathering of so many people

The Great Exhibition. This massive display of industrial machinery and goods inspired new confidence in the power of British manufacturing. *Credit:* Joseph Nash, Louise Haghe, and David Roberts, *Dickinson's Comprehensive Pictures of the Great Exhibition of 1851* (1852).

in one place—nearly 43,000 people visited per day, with a one-day record of 109,915—might trigger tumults or even revolts.

The new era of economic abundance and political instability soon gave rise to a revolution in economic thinking and a novel concept of scarcity. Neoclassical economics, as the new school of economics would become known, postulated that people in all ages and in all societies have insatiable desires for consumption, and that nature has always been limited yet is capable of improvement over time through scientific advances. One of the founders of neoclassical economics, William Stanley Jevons, had been inspired by what he observed firsthand as a fifteen-year-old during his many visits to the Crystal Palace Exhibition. Yet, once he turned to the study of economics in his mid-twenties, instead of grappling with abundance and economic growth, he focused on scarcity and infinitesimal changes. By maintaining a narrow focus on purely economic factors, Jevons and his fellow neoclassical economists removed attention from the ongoing political struggles for democracy and workers' rights. This analytical choice made the neoclassical idea of scarcity particularly popular among elites, who lived in constant fear of mounting pressures for radical social change.

Economic Abundance and Political Instability

Neoclassical economics emerged at a moment of massive economic upheaval. Earlier periods in history had witnessed their share of grandiose visions of progress, but it was only by the second half of the nineteenth century that science-based innovation truly transformed the world with a dizzying range of inventions. Scientists and entrepreneurs, aided by state institutions, introduced new productive technologies and refined old ones at a rapid pace. Large-scale and high-powered machines were made more efficient, user-friendly, and affordable, allowing great industrialists to transform nature on a scale never seen before. Soon after the First Industrial Revolution reached maturity, advancements in electrical and chemical engineering sparked the Second Industrial Revolution. The development of steam turbines, reliable cables, dynamos, and transformers allowed for the generation and transmission of electricity on a tremendous scale. Add the invention of the polyphase electric motor and humanity could now harness an enormous amount of energy for the mass production of goods, the lighting of entire cities, and the quick crossings of continents and oceans. Human muscular power had the capacity to generate up to ninety watts of energy, a large draft animal up to eight hundred watts, the waterwheel around four kilowatts, and the steam engine twenty kilowatts. By the 1860s, after the discovery of the first and second laws of thermodynamics, scientists were able to create steam turbines with the capacity to generate one megawatt.[2] Such command over energy opened up possibilities that were nothing short of astonishing. This period also witnessed the development of the gasoline-fueled internal combustion engine, which Karl Benz used to launch a commercial production of automobiles in 1886.

Access to large quantities of cheap energy, combined with Alfred Nobel's invention of dynamite in 1867 and the production of inexpensive, high-quality steel thanks to the Bessemer converter (1856), provided humanity with a capacity to control nature on a previously unimaginable scale. Railroad networks were built at a blistering pace, putting landlocked areas into communication with coastal cities, joining a global system of trade. The rapid displacement and genocide of indigenous populations, particularly in North America, made room for the development of large-scale agribusiness and new cities. Cheap steel made it possible to build more concentrated urban centers, with skyscrapers rising up toward the heavens, starting with Chicago's Home Insurance Building in 1884. Massive infrastructural developments, such as the London Tube (1863) and the Brooklyn Bridge (1869–1883), facilitated rapid urbanization and suburban sprawl. Mass-produced incandescent lightbulbs illuminated cities, most spectacularly in Paris, "the city of lights." Showing off their industrial power during the 1889 Exposition Universelle, the French built a 324-meter steel structure in Paris, complete with multiple lifts, at the top of which were two restaurants, a post office, and a viewing deck. To add to its luster, the Eiffel Tower was lit up by some ten thousand lights.

Cities could not have expanded at the rate they did had it not been for one of the most transformative chemical inventions of this era, the Haber-Bosch process of mass-producing fertilizers. In the years before the Great War, the German chemists Fritz Haber and Carl Bosch developed the method for converting nitrogen to ammonia, under extremely high temperature and pressure. By employing the technique of air liquefaction and manufacturing steel reaction vessels that could withstand the pressure, Haber and Bosch were able to produce synthetic fertilizer on an industrial scale. Without this invention, it is estimated that the world's population would not have been able to reach beyond 3.5 billion. Put another way, to produce the agricultural output levels achieved by the year 2000 would have required four times as much land.

Higher yields and vast increases in cultivated land—in particular, in the Americas—generated an enormous increase in food production. Steel-hulled ships powered by steam turbines and new cooling and canning technologies, enabled ordinary people in places like Edinburgh and Hamburg to enjoy beef from the Pampas and salmon from the St. Lawrence River. The fact that England, as the nineteenth-century world hegemon, had promoted trade liberalization since the 1820s further contributed to the dropping prices of foodstuffs. Lower prices for necessities increased the population's real wealth, which was already steadily inching upward thanks to higher real wages.

The technological revolution and the wealth it generated for both capital owners and wage earners gave rise to a new culture of consumption. The Crystal Palace Exhibition of 1851 had put the industrial marvels of the age on display; in the same spirit, the new bastion of consumption, the department store, invited the well-heeled to browse the new abundance of capitalism. While Le Bon Marché in Paris was the first *grand magasin* when it opened its doors to the public in 1852, closely followed by Harrods in London, the real golden age of department stores began in the 1880s. With their massive windows onto the street, airy, cathedral-like ambiance, and light- and mirror-filled interiors, they invited shoppers to participate in a spectacular alternate reality. The carefully curated environment was designed to make the consumer's imagination run wild. More than ever before, shopping had now become a bona fide leisure activity; even those visitors who were not actually buying anything were free to spend hours in the department store, fantasizing about beauty, style, and splendor. Middle-class women, empowered by their husbands' salaried positions as engineers, accountants, lawyers, and managers, were enticed to partake in the new extravagance. The department stores marketed themselves as providing an environment in which a female shopper could safely move around without the protection of a man. Guards were on hand, most of the clerks were women, and prices were fixed so the women did not have to engage in the unbecoming process of haggling. Department stores also provided places to dine and drink tea, to keep patrons in the building as long as possible. As the Western world entered the era of mass production, mass distribution, and mass consumption, consumers were faced with constant choices among a seeming infinitude of alternatives. The new consumer culture, with its promotion of insatiable wants, played a key role in the emergence of Neoclassical Scarcity.

With higher wages, increased purchasing power, more leisure, better education, and a freer press, the general population was also empowered to advocate for itself and participate more actively in politics. The waves of political unrest sweeping across Europe in 1848 had largely been brought under control by the authorities, but underlying discontent continued to fester during the second half of the nineteenth century. The disenfranchised were growing impatient with the failed promises of the

Enlightenment and the French Revolution. The vast industrial working classes were also starting to realize their collective power to shape the terms and conditions of their employment. Concurrent movements of socialists, communists, Marxists, and anarchists fed off each other, raising fears among the middle and upper classes. From the 1886 Trafalgar riots in London and the Haymarket riots in Chicago to the communist October Revolution in Russia in 1917 and the November Revolution in Germany (1918), industrial capitalism was under attack.[3]

Yet, as the legendary labor historian E. P. Thompson observed, working-class agitation in Britain did not cohere as a movement until the 1880s. The same can be said for many of the other European countries. It was in this decade that Belgium (1885), Switzerland (1888), Austria (1889), and Sweden (1889) formed socialist or social democratic parties; only the Social Democratic Party of Germany (1863) and Parti Ouvrier Français (1879) preceded them. When the four hundred delegates from twenty countries voted to form the Second International during a conference in Paris in the summer of 1889, the movement gained a global presence. They advocated for an eight-hour workday, the elimination of standing armies, and the use of mass strikes—a strategy that had been successfully employed by the London dockworkers in 1886 and the German miners in 1889. In England, the Socialist League was formed in 1885 by, among others, the poet and novelist William Morris (1834–1896), Karl Marx's daughter Eleanor Marx, and the latter's husband, Edward Eveling. As well as organizing street-corner rallies, Morris founded *The Commonweal,* one of many radical newspapers at the time, and used it to elaborate on Marxist ideas, the horrors of the factory system, and possibilities for socialism. In "How I Became a Socialist," Morris offered his definition of the term: "what I mean by Socialism is a condition of society in which there should be neither rich nor poor, neither master nor master's man, neither idle nor overworked, neither brain-sick brain workers, nor heart-sick hand workers, in a word, in which all men would be living in equality of condition, and would manage their affairs unwastefully."[4] Along with the formation of numerous labor unions, the 1880s also witnessed the launch of the Fabian Society. This non-revolutionary socialist society, in which the Irish playwright George Bernard Shaw would play an important role, paved the way for the Bloomsbury Group, in which John Maynard Keynes was a key member.

Europe thus experienced two distinct, yet interconnected, developments in the second half of the nineteenth century: a flourishing consumer culture and a variety of political movements for radical change. The former was steeped in aspirations of higher standards of living, the freedom that markets engendered, and rapid globalization—pursuits that served to strengthen the institutions of capitalism. The latter was motivated by dreams of increased equality, improved working conditions, and a restoration of community—goals that were in large part at odds with the emerging system of capitalism.

Neoclassical Scarcity

To counter the working-class movements and their radical ideologies, the bourgeoisie were in need of a new worldview. Liberal thinkers of the nineteenth century were charged with the task of rethinking capitalism as a set of opportunities and possibilities. Classical political economy was no longer of much use to them, as it had been distorted by the political agendas of Karl Marx, John Stuart Mill, and the Ricardian socialists. Marx had hijacked the labor theory of value to bring attention to the exploitation at the heart of capitalism and called for its immediate overthrow. Mill, while retaining faith in competition, had strongly advocated for a shift toward cooperative production and even self-identified as a socialist toward the end of his life. The Ricardian socialists were arguing that, since the labor theory of value defined value as rooted in human labor, workers ought to enjoy a much higher share of the firm's revenues. The socialist writer Thomas Hodgskin wrote about the workers as early as 1825, "The Labour is theirs, the produce ought to be theirs, and they alone ought to decide how much each deserves of the produce of all."[5] These three adaptations of classical political economy made the labor theory of value politically explosive and therefore of little use to those trying to marshal support for capitalism. As the prominent Swedish neoclassical economist Knut Wicksell would later note in a backward glance, "In the hands of the Socialists the [labor] theory of value became a terrible weapon against the existing order. It almost rendered all other criticism of society superfluous."[6]

Neoclassical economics gained popularity in the 1880s, yet it had been around since the early 1870s, with roots reaching back at least another two decades. Its originators found capitalism far from perfect, but firmly believed that with the proper guidance of economic theory it was possible to make capitalism better than any other known social system. By introducing a new way of conceptualizing wealth and shifting the main attention away from work and workers toward consumption and consumers—from pain to pleasure—it was possible to *prove,* as we discuss below, that no other economic system had the capacity to generate as much happiness and welfare as capitalism. They also insisted that their method, based on breaking down problems into the smallest possible components and applying mathematics to the analysis, elevated economics from a set of opinions to a science. In providing consumers and firms with utility and profit-maximizing strategies and liberal states with a powerful ideological defense against communism, this new understanding of capitalism proved remarkably successful, surviving to this very day.

Three thinkers articulated, independently of each other, the foundational ideas of modern neoclassical economics. While the exact terms in which they presented their theories differed, they agreed on the most fundamental principles. As a result of several nations having caught up with the pace of Britain's industrialization, the formation of modern economics was a polyglot endeavor, consisting of the Englishman William Stanley Jevons (1835–1882), the Austrian Carl Menger (1840–1921), and the Frenchman Léon Walras (1834–1910).

Jevons was the quintessential Victorian polymath, pursuing research and publishing in a variety of fields, including meteorology, geology, statistics, logic, and non-Euclidian geometry. He was intent on elevating economics into a science, which for him meant that the analysis should be conducted in terms of mathematics and mechanics.[7] He wanted to throw off "the incubus of bad logic and bad philosophy which [John Stuart] Mill's Works have laid upon us."[8] Jevons was also firmly opposed to both the labor and socialist movements, as they "arrest the true growth of our liberty, political and commercial." The average worker ought to be taught enough economics, he argued, to realize that "in the trade unions, in which he chiefly places his hope at present, there is no true individual freedom."[9]

Menger belonged to a group of Viennese economists, known as the Austrian School, who from the very start were intent on counteracting the socialist movement by providing a rigorous theoretical foundation for liberalism and arguing for the natural foundations of economic institutions.

The Vienna group was also committed to turning economics into a proper science. Although Menger did not express his ideas mathematically, he was drawing inspiration from the scientific and philosophical culture flourishing in Vienna at the time.[10]

Walras, the son of one of the progenitors of the new school of economics, did not turn to economics right away. He first trained as an engineer, but never earned a degree. Instead, he wrote a novel and tried his hand at philosophy, before he took up employment as a bank manager, journalist, and railway clerk. Even more than Jevons, he was intent on systematizing economics and turning it into a mathematical science on par with physics and mechanics. He was influenced by his father's conviction that the land belonged to everyone and that landed proprietors therefore should not be able to live off unearned rents. The rent-revenues should instead go to the state, which would greatly benefit the community and obviate the need for taxes. Yet, despite embracing this Saint-Simonian idea of state-owned landed property, he certainly was not opposed to private property and markets in any other domain of society. Indeed, his general equilibrium analysis played an important role in promoting the argument that market-societies were best equipped to address the economic challenges facing humanity. Walras unequivocally declared, "Freedom of exchange! Laissez demander, *laissez offrir. Laissez produire, laissez entrer;* or, to go back to an excellent Physiocratic formula: *Laissez faire; laissez passer.*"[11]

At the heart of the neoclassical notion of scarcity was a fundamental rethinking of the origins of value. While the labor theory of value articulated as early as William Petty, and all the way through to John Stuart Mill, deemed labor the source of all value. The neoclassical economists made people's subjective valuations—how much happiness they derived from a commodity—the determinant of value. Value, Menger explained, is "nothing inherent in goods, no property of them, nor an independent thing existing by itself. It is a judgement economizing men make about the importance of the goods at their disposal for the maintenance of their lives and well-being."[12] Jevons similarly, though more succinctly, declared that "value depends entirely upon utility."[13] Since happiness, or utility, was difficult to measure, a person's willingness to pay for a good was used as a proxy. The next pivotal intervention the new school of economics introduced, which would also give them their name, *marginalists,* was to focus not on the total amount of utility derived from consumption, but rather on the utility from the marginal unit. That is, the value of a good was determined by people's utility and their willingness to pay for an additional unit.

The marginalists argued that utility tends to decrease as a person consumes additional units—a concept that was later labeled *diminishing marginal utility*. While the consumption of the first unit generated a lot of happiness, the consumption of the second unit, while still positive, yielded a bit less utility than the first one, and so on. The hypothetical process continued until finally the marginal utility reached zero—the exact moment of which would be determined by the person's psychology and the type of commodity. This meant that under normal circumstances people were willing to pay a lot for an additional unit of a thing that only exists in small quantities, while they were not willing to pay much at all for an additional good or service that exists in abundance.[14]

The marginalists theorized humanity as engaged in the maximization of utility and the minimization of disutility. Without engaging in extensive empirical research on consumer behavior or delving deeply into the psychological literature, Jevons suggested that "Economics is a Calculus of Pleasure and Pain."[15] The marginalists argued that people made consumption choices on the basis of what commodities contributed most to their overall individual happiness. This meant that the first good they bought was the one to which they assigned the highest marginal utility. They continued buying this good until the marginal utility fell below that of the next most desirable good. At this point, they selected the commodity that now had the highest marginal utility and so on and so forth. Eventually, the utility-maximizing person ended up with a basket of goods in which all goods had the same marginal utility. This is the condition of optimality. If the condition did not hold, it meant that the person could have achieved a higher total utility if they had consumed more of the good with the higher marginal utility

To obtain their optimal mix of goods, people had to trade with other people, who were themselves engaged in the process of maximizing their utility. The rate at which people were willing to exchange a good was determined by the marginal utility they assigned to it, which, by extension, meant that the resulting equilibrium price was proportional to people's relative desires. At such an equilibrium price, people had acquired what for them was the most desirable mix of goods, given their wealth. When generalized in a system of simultaneous equations, the theory, Walras showed, maintained that as people in market societies optimize their utility they inadvertently achieve the socially optimal allocation of goods.[16] This was a remarkably powerful result.

Neoclassical economics thus assumed that people had an insatiable desire, not for any specific good, but for consumption in general. There was no point of satiation since there was always something more to consume. This unavoidable condition of insatiability was assumed to be "imbedded in our nature," acting as the "great spring of human action" upon which all economic behavior was based.[17] Scarcity thus constituted a reasonable starting point for all economic inquiry; it was to economics what velocity and heat were to mechanics and physics.[18]

Scarcity was everywhere. Wants always outstripped means. According to neoclassical economics, it did not matter if a person were rich or poor, young or old. As Menger explained, "one man may live in a palace, consume the choicest foods, and dress in the most costly garments. Another may find his resting place in the dark corner of a miserable hut, feed on leftovers, and cover himself with rags. But each of them must try to satisfy this needs for shelter and clothing as well as his need for food."[19] While the poor person's experience of scarcity is easy to imagine, the rich man also faced real trade-offs, as "the needs of men can display an almost unlimited variety."[20] Because tastes were ever evolving and there was no end to the variety of shelter, food, dress, and leisure activities available in the new world of capitalist abundance, there was always something else that people could enjoy. Jevons wrote:

> The necessaries of life are so few and simple, that a man is soon satisfied in regard to these, and desires to extend his range of enjoyment. His first object is to vary his food; but there soon arises the desire of variety and elegance in dress; and to this succeeds the desire to build, to ornament, and to furnish— tastes which, where they exist, are absolutely *insatiable,* and seem to increase with every improvement in civilisation.[21]

A consumer did not just view things he or she did not have as scarce. Objects already in the consumer's possession were also treated as scarce, even if he or she had no interest in consuming them. The rationale was that they could always be used to trade for something else. Walras argued that appropriating any good capable of satisfying someone's wants yielded "a double advantage: not only do they assure themselves of a supply which can be reserved for their own use and satisfaction; but, if they are unwilling or unable to consume all of their original supply themselves, they are also in a position to exchange the unwanted remainder for other scarce utilities which they do care to consume."[22] Because the market makes goods fungible, Menger insisted, every resource or commodity was scarce; no portion could be given up without causing a sacrifice of want-satisfaction. Humanity thus lived in a world characterized by a perennial condition of scarcity.

Questioning the Neoclassical Notion of Insatiability

The driving force behind Neoclassical Scarcity was people's ever-expanding desires for consumption. It was these desires that dictated what and how much was produced in society. Instead of thinking about production as dictated by firms or, in the Marxist tradition, bloodthirsty capitalists, neoclassical economists insisted that everything that happened in the economy was ultimately dictated by what the consumers wanted—a concept referred to as *consumer sovereignty*. However, not everyone accepted the notion that consumer demand ought to determine what and how much were produced in a society, nor did they believe that modern consumption constituted the foundation for happiness and that it was beneficial to society in general. As such, they questioned the very foundation of the neoclassical version of scarcity.

The eccentric yet brilliant Norwegian-American social theorist Thorstein Veblen (1857–1929) challenged the neoclassical—a term he coined—notion that consumption promoted happiness and well-being. Veblen was trained in economics, philosophy, and sociology. As an undergraduate student, Veblen studied with a young John Bates Clark, who would later go on to become one of the most celebrated neoclassical economists in the United States, and as a graduate student he read philosophy with the founder of American Pragmatism, Charles Sanders Peirce. Veblen argued in his classic book *The Theory of the Leisure Class* (1899) that much consumption—in particular, the luxury consumption of the well-to-do—

was motivated not by genuine needs or wants but by the status it conferred on the consumer. He saw high-end consumption as a competitive game, in which people tried to upstage each other by conspicuously displaying their purchases. The competition was not just crudely about who could consume the most expensive clothes or house, but also about who exhibited the most refined tastes. This required evidence of gentility and good breeding, which required a tremendous amount of time and education.[23] A person had to be able to discern between "the noble and the ignoble in consumable goods" to become "a connoisseur in creditable viands of various degrees of merit, in manly beverages and trinkets, in seemly apparel and architecture, in weapons, games, dancers, and the narcotics."[24] To Veblen, this was sheer waste. Consumption, at least the conspicuous sort, had little to do with genuine welfare and everything to do with status and power. People might always want more, not because the goods made them happy, but because of the satisfaction that came from observing the admiring glance of others. The desire for distinction, to Veblen, was not a legitimate desire, and fulfillment of which society ought not to be organized around it.

The absurdity of allowing consumer desires to dictate what and how much is produced in a society was the topic of some of the most iconic fin de siècle novels. For example, Joris-Karl Huysmans (1848–1907) in *Against Nature* (1884) depicted a decadent and self-indulgent aristocrat who isolated himself in a villa to cultivate his tastes and to enjoy the full bounty of modern luxuries.[25] Everything this connoisseur bought, he carefully contemplated, scrutinizing its colors, fabrics, materials, feel, country of origin, and so forth. The more refined, rare, voluptuous, and luxurious the object was, the more he valued it. In one of the most decadent scenes, he finds himself reflecting on the fact that the tint on the shell of his newly acquired tortoise does not properly match the sheen of his new carpet. "At last he came to the conclusion that his original idea of using a dark object moving to and fro to stir up the fires within the woolen pile was mistaken." The carpet was "too bright, too garish, too new looking." He had to find a way to mute the colors in the carpet. His solution was to have his tortoise's buckler glazed in gold."[26] But this did not suffice—the tortoise needed something even more brilliant—so he decided to encrust its shell with precious stones. He considered all the available precious stones, but refused them one by one, as they had become vulgar. Businessmen, small shop owners, and even butchers' wives had begun wearing them.

Finally, he came upon a sapphire: "Alone among these stones, the sapphire has kept its fires inviolate, unsullied by contact with commercial and financial stupidity."[27] Yet he could not quite commit to the sapphire, either, as he did not appreciate how it refracted the light from the incandescent lightbulbs. In the end, he settled on a combination of gems, and had his jeweler arrange them as a beautiful bouquet of flowers on the back of his tortoise. And so the book continues, with Huysmans describing the dandy figure striving to arrange his life according to his own highly refined aesthetic, which would likely have struck the reader as the self-indulgence of an insatiable hedonist. Huysmans's protagonist, the product of the modern cosmopolitan consumer culture, was the epitome of the insalubrious marriage between modern excess and meaninglessness, between modern prosperous capitalism and an increasingly subjugated nature.

While Huysmans criticized the absurdity of an economy dedicated to the satisfaction of irrational consumer desires, the French novelist Émile Zola (1840–1902) offered a cultural critique of modern consumerism. After patiently conducting research in the department stores of Paris, he wrote *Au Bonheur des Dames* (1883), which focused on the negative impact of consumption on women. While certain men, fops and dandies, were seen as liable to the allure of shopping, middle-class women were perceived as particularly susceptible to the temptations of fashion, luxury, and flare—exactly what the department stores offered. Women were characterized as easily seduced by a ruthless marketing machinery preying on their fickle desires. Embedded in the department store's carefully staged marketing theatrics, Zola's main character, Madame de Boves, found herself overwhelmed by the temptations and ended up shoplifting even though she had money in her pocket. Zola wrote in his notebook that "Women go [to the department store] to pass the time, just as they used to go to church: it is . . . a place where they become impassioned, where they enter into a struggle between their passion for fashion and their husband's budgets."[28]

Zola captured a number of related anxieties about the new culture of consumption, to which none of the neoclassical economists paid much attention. He expressed the prevailing societal fear that the new havens of consumption compromised bourgeois women's integrity, self-command, piousness, marital commitment, reason, civic virtue, respectable sociability, and observance of property rights.[29] The programmed female consumer thus served as an agent of social unrest and a threat to polite society. Nostalgic dreams of traditional forms of community, hierarchy, honor, and respect were shattered by marketplace individualism and its hordes of desirous women. Incessant consumption—the underlying force behind scarcity—was thus seen as a vulgar and inauthentic practice, not a source of genuine happiness.

Veblen's contemporary the French economist Charles Gide (1847–1932), uncle of the prominent novelist André Gide, challenged the neoclassical understanding of consumption from yet another vantage point. As a Protestant Christian socialist, he found it unacceptable that the ends of human economic activity were not subject to any type of normative assessment by economists. The ends people pursued were not the concern of economists; they could be noble or base, material or immaterial, it simply did not matter. Walras had offered a particularly blunt statement about the irrelevance of normativity: "whether a drug is wanted by a doctor to cure a patient, or by a murderer to kill his family is a very serious matter, but from [the economist's] point of view, it is totally irrelevant."[30] In the spirit of John Ruskin's notion of "illth," the negative moral consequences of consumerism, Gide suggested that consumers ought to be held responsible for the social impact of their purchases. If they were not, it would foster an ethical nihilism. Gide insisted that every consumer ought to exhibit solidarity toward their fellow citizens and take into account the effects their consumption had on other people and on society in general. Buyers should inform themselves about the conditions in which the products they bought were produced—did the companies pay their workers decent wages, were safe-labor practices followed, and did they uphold high standards of workmanship? To help consumers make ethical choices and to empower them to have an impact on the community, Gide promoted the formation of consumer cooperatives. These would establish a "reign of the consumer" to rein in the abuses staining capitalism.[31] Gide did not see anything socially optimal about utility-maximizing individuals interacting as monads in the marketplace. To theorize the economy as such only served to further naturalize unchecked consumerism.

Veblen, Huysmans, Zola, and Gide confronted the neoclassical assumption that human desires for consumption were intrinsically insatiable. They dispelled the notion that the satisfaction of modern consumer desires promoted happiness and that consumer demand emerged organically from a person's intrinsic psychology. More recent scholarship has argued that desires for consumption are always shaped by a host of social and cultural circumstances. Only at a stage of history in which there was a near-infinitude of consumables did it make sense to attribute insatiable desires to the general population and thus to think of the world as ruled by a perpetual condition of scarcity. If goods could not be transmuted into other goods through the market, people would not treat everything around them as scarce. In the absence of a cosmopolitan consumer culture, moreover, it was difficult to envision how tastes could have become as dynamic and fluid, as this had not been the general condition of humanity through history.[32]

The Prospects of Overcoming Scarcity

Alfred Marshall, the most influential of the second-generation marginalists, at least in the anglophone world, put forth the idea that Neoclassical Scarcity could be ameliorated, if not entirely overcome, over time.[33] Marshall argued that it was possible to calm people's desires by educating the working classes and instilling a sense of economic chivalry into the business elites. Trained as a mathematician at the University of Cambridge, Marshall became famous as a result of the publication of his *Principles of Economics.* Mostly written in the 1870s, though not published until 1890, his book became the standard textbook in economics for decades. Marshall offered a synthetic account of the new economics and also made numerous important contributions and clarifications on his own. But Marshall was not solely concerned about the world of high-mathematical theory, he was genuinely curious about the actualities of capitalism. He visited mines, workshops, and factories throughout Britain where he interviewed business owners, industrialists, merchants, engineers, trade-union leaders, and shop-floor workers. Right around the time he was writing the *Principles,* he used an inheritance from an uncle to tour the United States, saying he "wanted to see the history of the future in America."[34] While he met with dignitaries, such as Ralph Waldo Emerson, and visited communes run by disciples of Robert Owen, most of his time was spent observing the American economic wonder. He traveled on the rapidly expanding rail network, observed the frenetic endeavors to build new cities, and visited factories producing everything from machines to pianos. He was impressed by how capital and technology "enable us to make nature work for us as an obedient and efficient servant."[35] Despite some earlier flirtations with Romantic socialism of the kind William Morris promoted, which he abandoned when he began to fear that socialists were too enamored with violence, Marshall firmly believed that no other system was as capable as capitalism to address the main challenges facing humanity, which included poverty, inequality, and ignorance, and that no other brand of economics was better suited than neoclassical economics to explain and manage capitalism.

Marshall embraced the same notion of scarcity as his predecessors. "Human wants and desires," he wrote, "are countless in number and very various in kind." While the desire for one particular good can be satisfied completely, at "every step in his progress upwards increases the variety of his needs together with the variety in his methods of satisfying them. He desires not merely larger *quantities* of the things he has been accustomed to consume, but better qualities of those things; he desires a greater choice of things, and things that will satisfy new wants growing up in him."[36] Not unlike Veblen, Marshall acknowledged that the "desire for distinction," along with a desire for variety, powerfully shaped people's preferences. Indeed, the "efforts to obtain distinction by dress are extending themselves throughout the lower grades of English society."[37] This, Marshall considered, was a favorable development; it was a sign of progress that the poor were joining the middling sorts.

Marshall was optimistic that it was possible to eliminate the most egregious features of the capitalist class system. He questioned whether "it is necessary that there should be any so-called 'lower classes' at all: that is, whether there need be large numbers of people doomed from their birth to hard work in order to provide for others the requisites of a refined and cultural life, while they themselves are prevented by their poverty and toil from having any share or part in that life."[38] To emancipate workers, bring them out of poverty, and give them an opportunity to exercise their intellects, one must provide them with better education. First, a better education was the best way to increase wages for the poor. Shattering the old notion of a fixed wages fund (a fixed limit on how much each firm can allocate toward wages) and the iron law of wages (the tendency for wages to approach the bare minimum standard of living), Marshall argued that a better educated and more productive working class would earn higher wages. According to the model of perfect competition, neoclassical theory maintained that wages were equal to the marginal revenue product of labor, which meant that if the marginal product of labor went up, so would wages. Second, a better educated workforce would be able to operate more sophisticated technology. This would rescue the workers from the most stultifying manual labor and engage them in tasks, in collaboration with other skilled workers, that were intellectually rewarding. This process would continue "till the official distinction between working man and gentleman has passed away; till, by occupation at least, every man is a gentleman."[39] Third, with more time on their hands and with better education, the workers would also be able to develop a wider "range of pleasures" and purge themselves of their "desire for coarse delights."[40] Based on his extensive travels and numerous interviews, he concluded that "All ranks of society are rising; on the whole they are better and more cultivated than their forefathers were," adding that, with "the aid of education," the people's "moral strength is gaining new life."[41]

While a greater material abundance and a more refined working class constituted a good start to the easing of scarcity, Marshall insisted that business elites also had to rein in their consumption. He called for men of business to limit their preoccupation with pecuniary gain and conspicuous consumption. While they should no doubt continue to invent, innovate, and employ their resources as efficiently as possible to earn a profit, they should curtail their ingrained pursuit of personal enrichment. Great amounts of wealth accumulated in the hands of the few "contribute very little towards social progress." Once again echoing Veblen, Marshall suggested that the widespread pattern of conspicuous consumption must be "regarded as socially wasteful."[42] He proposed instead that business men should be compensated in a less frivolous manner. He wrote:

> if society could award this honour, position, and influence by methods less blind and less wasteful; and if it could at the same time maintain all that stimulus which the free enterprise of the strongest business men derives from present conditions, then the resources thus set free would open up to the mass of the people new possibilities of a higher life, and of larger and more varied intellectual and artistic activities.[43]

For businessmen to adopt this new ethic, they had to embrace a spirit of "economic chivalry." Indeed, he insisted that until economic chivalry was embraced by the business community, "the world under free enterprise will fall short of the finest ideals."[44]

Over time, the chivalrous business elite in collaboration with a better educated working class would ameliorate the condition of scarcity and pave the way for new kinds of progress. Not unlike David Hume, Marshall believed that it was possible for humanity to evolve in ways that changed the nature of people's desires and sensibilities, thereby removing vulgar selfishness and mindless hedonism as the source of scarcity.

After teaching many years at the University of Bristol, Marshall returned to the University of Cambridge, where he became its first professor of political economy. One of his most brilliant students was John Maynard Keynes, whom Marshall allegedly persuaded to study economics rather than philosophy. Keynes, celebrated most widely for his role in the formation of the field of macroeconomics, went on to have an illustrious career in both academia and public policy. He not only embraced the bulk of Marshallian economics, he also seems to have been inspired by Marshall's vision for an end to scarcity.

Keynes denied the neoclassical assumption that scarcity was perpetual; instead he insisted that there was a way to escape it altogether. Drawing on Veblen and perhaps even Nicholas Barbon (see Chapter 2), Keynes divided desires for goods into two categories: "those needs which are absolute in the sense that we feel them whatever the situation of our fellow human beings may be, and those which are relative in the sense that we feel them only if their satisfaction lifts us above, makes us feel superior to, our fellows."[45] While he acknowledged that desires for the second set of goods "may indeed be insatiable," this was certainly not the case for the former. He projected that within the next hundred years, by around 2030, a point would be reached when basic "needs are satisfied in the sense that we prefer to devote our further energies to noneconomic purposes."[46] When this milestone was reached, the whole population would be assured of adequate housing, clothing, nourishment, and comforts. The "economic problem" facing humanity would thus not remain "the permanent problem of the human race" and people could stop obsessing about money and economics. Given that the "pseudo-moral principles which have hagridden us for two hundred years," cannot be entirely purged in an instance, some people will continue to engage in accumulation and conspicuous consumption. But with time, such "semi-pathological propensities" would eventually evaporate.[47]

In separating wants that are infinite, relative, and insatiable from needs that are finite, absolute, and satiable, and thereby opening up the possibility that scarcity could one day be abolished, Keynes offered a more radical vision of the future than Marshall. Once the struggle to satisfy basic human needs had been overcome, people would turn to the cultivation of the "art of living," a term he borrowed from John Stuart Mill (see Chapter 4). In Mill's vision for the future there would be plenty of scope for "mental culture, and moral and social progress; as much room for improving the Art of Living; and much more likelihood of its being improved, when minds ceased to be engrossed by the art of getting on."[48] Keynes was similarly optimistic, although he acknowledged that people had to endure a period of "dread and adjustment," as they rid themselves of the habits "bred into [them] for countless generations."[49] As people embraced the challenge to cultivate themselves "into a fuller perfection, the art of life itself," they would "be able to enjoy the abundance when it comes."[50] With only three hours of labor necessary per day—compared to Thomas More's Utopians who labored six or Godwin's farmers who worked only a single hour—Keynes envisioned that humanity would soon develop higher enjoyments and learn "how to pluck the hour and the day virtuously and well."[51]

Marshall and Keynes's vision for a future attenuation or end to scarcity had little or no impact on modern economics—some may even suggest that Keynes himself never took the idea seriously and only wanted to instill hope in the people who were suffering through the worst of the Great Depression.[52] While the critique of consumerism continued in sociology, philosophy, and heterodox economics, neoclassical economists doubled down on their core method and principles. They would confidently promote the idea, formulated by Jevons, that neoclassical economics applied "more or less completely, to all human beings of whom we have any knowledge."[53]

The Proof (or Illusion) of Optimality

The new generation of neoclassical economists taking the world by storm in the 1950s proclaimed that they had found a way to prove mathematically that there existed an optimal relationship between the economy and nature. In their 1954 article, Nobel Prize winners Kenneth Arrow and

Gérard Debreu showed that free people and firms rationally deciding on the use of their time and resources with the aim of maximizing their utility and profits, without paying attention to anyone else and without any outside interference, would reach an equilibrium.[54] This resting point constituted the best possible outcome, given society's resources and technology and people's preferences. The fact that market societies were capable of establishing such an optimal solution suggested, according to the neoclassical economists, that capitalism was efficient, fair, and optimal.

The basic neoclassical model employed three key assumptions about individuals and firms. First, people were rational, meaning that they were able to rank the ends available to them. Second, people had full information, in the sense that both buyers and sellers were fully informed about prices, costs, and quality. Third, every market was perfectly competitive, which meant that no individual person or firm had the power to influence prices. In addition to the utility-maximizing consumption decisions we discussed above, neoclassical economists extended their marginal analysis to other pivotal economic decisions. For example, in deciding how much to work, people compared the marginal benefits (the wage) to the marginal cost (the disutility of work) of working one more hour or one more day. On the other side of the labor market transaction, profit-maximizing firms evaluated the marginal benefit of hiring one more worker (the marginal revenue product of labor) to the marginal cost (the wage). The labor market eventually found its equilibrium, and because it was the result of everyone's maximizing choices, all participants were satisfied with the outcome. Moreover, a sense of justice was established, in that every worker was paid their marginal revenue product, which was equal to what they contributed to the productive endeavor. The other two factors of production, land and capital, were also paid their marginal revenue product, which ensured that all owners of the factors of production were properly and meritocratically compensated. There was no exploitation, no coercion, and no abuse of power in this neoclassical world.

Lastly, according to these theories, firms optimized their decisions regarding how much output to produce. The decision-rule each firm followed was to produce additional units as long as the marginal revenue (price) was greater than the marginal cost. Because most production processes were characterized by diminishing marginal returns, after a certain level of output the marginal cost began to gradually rise. This meant that eventually the marginal cost would become equal to the fixed marginal revenue (recall that all firms treat prices as fixed under conditions of perfect competition). A profit-maximizing firm thus ended up producing the output at which the last unit contributed as much to revenues as it did to costs. Since the price was ultimately determined by the intensity of people's desires, the equilibrium level of output could be said to have occurred when marginal utility was equal to the marginal cost. Phrased differently, each firm's optimal production strategy ensured that each sector produced the *socially* optimal level of output.

In mathematically proving the existence of an equilibrium, Arrow and Debreu formalized Adam Smith's *invisible hand* or what neoclassical economists called the First and Second Welfare Theorems. The first theorem stated that the equilibrium allocation was Pareto optimal, which meant that no further exchanges could be carried out that would improve the well-being of one person without making anyone else worse off. The second theorem declared that any Pareto optimal allocation could be supported by a competitive equilibrium, which implied that there was no need for any government redistribution of resources to achieve the ideal outcome.[55] Together, these theorems established that unregulated markets ensured that the socially optimal level of output was produced. Markets thus provided the best path to the optimal relationship between economy and nature.

This claim served powerful ideological interests in the Western world during the Cold War. Incidentally or not, the RAND Corporation, a think tank established in 1948 to provide scientific and social scientific research to the US armed services, played an important role in promoting this line of research. While it remains controversial among historians whether the theory was ideologically motivated, its ideological utility during the Cold War is uncontested.[56]

Friedrich Hayek (1899–1992), a disciple of Menger and the founder of the conservative Mount Pelerin Society, employed neoclassical theory to directly challenge the feasibility and desirability of communism. In his bestselling book *The Road to Serfdom* (1944), he argued that unobstructed and unregulated markets generated a set of prices that provided actors in the economy with all the requisite information they needed to allocate their resources in a manner that produced the social optimum.

Together with Milton Friedman, Hayek created a powerful presence of free-market, anti-government-intervention economists at the University of Chicago. One of their colleagues, Gary Becker (1930–2014), extended the results of neoclassical theory from the markets for consumer goods and factors of production to human action more broadly, including marriage, crime, and discrimination. Since everything people wanted, Becker reasoned, was ruled by scarcity, people found themselves constantly economizing, which meant that the mathematical tools of optimization were relevant to the study of nearly all human behavior. The idea that one might derive optimal behavior in crime, marriage, and discrimination energized economists, but baffled other scholars, including Michel Foucault, who engaged Becker's work in his lectures on biopolitics. Hayek, Friedman, and Becker would each go on to win the Prize in Economic Science in Memory of Alfred Nobel, in 1974, 1976, and 1992, respectively.[57]

The power to declare what constituted the optimal allocation of resources was also a useful stratagem against environmentalists, whom the conservative think tank The Heritage Foundation named the "greatest single threat to the American economy."[58] Ever since late-nineteenth-century conservationists and preservationists began raising awareness of how the "efficient" exploitation of the environment put humanity in peril, there had been growing anxiety in some circles that free-market societies were producing at unsustainable levels, far beyond what was optimal from the point of view of the environment. Critics argued that the neoclassical optimality condition was vacuous at best, in part because demand did not capture society's collective desires, only desires backed by purchasing power, and in part because far from all costs incurred in the production of goods were paid for by the producing firms. Because firms did not pay for all the costs associated with their production, in particular those imposed on the environment, such as smog, acidification of oceans, species extinction, and global warming, the profit-maximizing level of output for firms was too high. That is, while the firms' decisions were still optimal for them and their consumers, they were no longer compatible with the stability of the environment. To address this concern, a number of neoclassical practitioners dedicated themselves to the development of a solution whereby firms were forced to internalize the relevant costs, the so-called externalities. They ended up going back to the pioneering work of A. C. Pigou and his *Economics of Welfare* (1920) for inspiration.[59] Pigou, Marshall's successor to the chair of political economy at the University of Cambridge, had suggested that one way to solve this problem was to levy a tax—of the same magnitude as the firm's externalities—on the producer, which would increase its marginal cost and thus lower its profit-maximizing level of output. If the externalities could be accurately quantified, which is an entirely different can of worms, then the First and Second Welfare Theorems would once again hold.[60]

Not all neoclassical economists were pleased with this solution, in large part because it required government intervention, but also because it was difficult to implement in practice. Instead, they promoted an alternative. Ronald Coase, a University of Chicago Nobel Prize winner, developed what would become known as the Coase Theorem, which proclaimed that the problem of externalities was not just about making sure that firms internalize social costs, but also about the *manner* in which they were implemented.[61] This was, first and foremost, a problem of legal rights, more precisely, property rights, according to Coase.[62] He specified that the proper way to think about externalities was to think of it as a reciprocal problem. He used the example of a chemical company discharging toxic waste into a river, which poisoned the fish and thereby put an end to fishing. While the "emotional" response to this problem was to get rid of the industrial waste and clean up the river so that the fish caught would be safe enough to eat, the "rational" solution was to figure out whether the value of the chemicals produced by the firm was greater than the value of the fish. "What we really want," he wrote, "is the right amount of pollution—the amount, that is, which maximizes the value of production."[63] The most efficient solution to this problem was to ensure that all the relevant resources were privately owned. If transaction costs were low enough, the owners would then be able to negotiate and thus arrive at the socially optimal level of output and pollution.[64] No government intervention was necessary; the market would organically achieve the optimal allocation. This argument was more convincing when pollution was local. However, Coase's model was much more problematic when applied to oceanic and atmospheric pollution, as strict private property rights cannot be assigned. While some of Coase's followers tried to make this extension, Coase himself never spoke in such terms.[65]

Neoclassical Cornucopianism

The neoclassical general equilibrium analysis was situated within a static framework, yet there was an underlying sense that the economy was on an upward trajectory. At any given moment, subject to fixed preferences, technology, and resources, firms and individuals made choices that maximized their utility and profits. Each moment yielded an optimal allocation of resources—a best way for humanity to use nature. Although the analysis was static and focused on marginal change, there was an underlying assumption that each static moment linked up to a dynamic cornucopian process. While their formal models did not capture their faith in progress, most neoclassical economists embraced the spirit of growth at the heart of capitalism. Alfred Marshall, for example, noted, "As civilization has progressed, man has always been developing new wants, and new and more expensive ways of gratifying them. The rate of progress has sometimes been slow, and occasionally there has even been a great retrograde movement; but now we are moving on at a rapid pace that grows quicker every year."[66] Marshall thus saw the economizing behavior that generated the static equilibrium at any given moment in time as part or broader secular trend of economic growth. He added, in defiance of John Stuart Mill, "There seems to be no good reason for believing that we are anywhere near a stationary state in which there will be no new important wants to be satisfied; in which there will be no more room for profitably investing present effort in providing for the future, and in which the accumulation of wealth will cease to have any reward."[67]

The depletion of natural resources also would not put a strain on continued economic advancement. In the spirit of Francis Bacon, neoclassical economists considered nature a standing reserve, a set of resources existing exclusively for the use of humanity. The point was to maximize consumption by making use of available resources—there was no reason to hold back, either on ethical or preservationist grounds—so that the welfare of consumers could be optimized. Nature entered the neoclassical models only to the extent that it imposed costs on the production of goods. These costs were seen to reflect the degree to which the resources were in short supply. The rarer they were, the more expensive they were to the firm and, in extension, to the consumers, which meant that people would demand less. If a resource was in danger of running out completely, the price would skyrocket, which would create market signals for science to develop substitutes that would enable the economy to keep on going. Markets and science thus ensured that the depletion of resources would not put a stop to economic progress.

To understand change over time, the marginalist economists compared how the optimal resource allocation changed when any of the variables held fixed were altered, an analysis they referred to as comparative statics. However, this did not amount to a real theory of why or how these variables actually expanded over time. In short, neoclassical economics lacked a theory of economic growth. This became particularly problematic in the aftermath of World War II, when economic growth emerged as the perceived panacea, both in the East and in the West. Nikita Khrushchev, for example, proclaimed in 1958 that "Growth of industrial and agricultural production is the battering ram with which we shall smash the capitalist system."[68] The United States and its allies in Western Europe similarly viewed economic growth as the main weapon whereby they would prevail in the Cold War, as well as the key to winning the hearts and minds of the Global South through the process of development.[69] Cornucopian growth was also regarded as the foundation for the emerging welfare state, the defusing of class tensions, and the achievement of a higher standard of living. Growth was thus the key to capitalism's triumph.

Aided by an ambitious endeavor to collect national income statistics, begun during World War II, MIT economist Robert Solow (1924–) developed a neoclassical growth theory in the 1950s. The key factors that determined a nation's rate of growth, the 1987 Nobel Prize winner argued, were labor, capital, and technological progress. Population growth dictated the rate at which labor input increased, and the savings rate and the rate of depreciation determined the rate at which capital grew. While increases in these factors of production were important, Solow found that improvements in technology accounted for four-fifths of economic growth. The rate at which the economy grew was thus mostly determined by the development of knowledge. But what about nature and natural resources? Did they have an impact on economic growth? Did the relationship between the economy and nature play a role in his thinking about growth? The simple answer was no. Natural resources and the environment were absent in Solow's model; abundance of natural resources did not cause growth, nor did their dearth hamper it. He noted rather flippantly in a 1974 essay that the world "has been exhausting its resources since the first cave-man chipped a flint, and I imagine the process will go on for a long, long time." What made nature largely irrelevant to the question of growth was the fact that human ingenuity had proven so strong that people had managed to develop substitutes whenever they ran out of energy or re- sources. "The world can, in effect," he argued, "get along without natural resources, so exhaustion is just an event, not a catastrophe."[70] Solow had no interest in entertaining the concerns of the environmentalist move- ment, nor did his growth model have much to say about the relationship between nature and economy.

Solow's growth model remained dominant until the early 1990s, when models of economic growth were introduced by the University of Chicago's Robert Lucas and Chicago-trained Paul Romer, respectively, in which human capital improvement and technological progress were treated as en- dogenous. That is, not only did these factors cause economic growth, eco- nomic growth further contributed to the advancement of human capital and technological improvement. By defining knowledge as having increasing re- turns to scale, Nobel laureates Lucas and Romer theorized a mutually rein- forcing relationship between economic growth and knowledge—a virtuous circle. As long as people remained creative and innovative, there would be no end to economic growth. These endogenous-growth models brought eco- nomics back to the original Cornucopian worldview of the seventeenth- century alchemical thinkers (see Chapter 2). For them, nature was an infi- nite treasure, the fruits of which could be continuously harvested as long as people kept on cultivating their imagination and promoted the advancement of scientific knowledge. With a never-ending series of improvements in knowledge, all environmental obstacles to economic growth would be tran- scended, and the economy could continue to grow, *ad infinitum.*

In the 1970s, as the environmental movement gathered momentum, neoclassical economists not only dusted off Pigou's old model on exter- nalities, as discussed above, but also revised their growth theories to incorporate some environmental considerations. One of Solow's many pro- tégés, William Nordhaus (1941–), who would later go on to share the No- bel with Romer in 2018, asked in a 1972 article (coauthored with James Tobin, yet another Nobel laureate): "will natural resources become an increasingly severe drag on economic growth?" Their answer: "We have not found evidence to support this fear. Indeed, the opposite appears to be more likely: Growth of output per capita will accelerate ever so slightly even as stocks of natural resources decline."[71] Indeed, Nordhaus argued, similar to Solow, that "land and resources" have rightly been "dropped" from consideration because "reproducible capital is a near-perfect substitute for land and other exhaustible resources."[72] While people and politicians should not worry about the exhaustion of natural resources, they should do something about pollution. In the spirit of Pigou, Nordhaus suggested that firms should be forced to pay for their emissions, the effect of which, neoclassical economic theory predicted, would be a reduction in output and pollution. He did not want the penalties to be severe, however, as this would reduce the overall welfare of the present population. If economic growth were reduced today, he believed, chances of addressing problems in the future would worsen. Instead, he encouraged his readers to remain optimistic about the capacity of science to solve the problem.

Nordhaus, developing his thinking further, went on to formulate what he called his Dynamic Integrated Model of Climate and the Economy (DICE). This model estimates the social cost of the emission of carbon by calculating the sum total of future harms anticipated from additional emissions and translating that to its net present value. The intent was to figure out the optimal carbon tax over time and how much ought to be spent on mitigation efforts.[73] Nordhaus suggested that aggressive spending on mitigation efforts in the immediate future was potentially worse than doing nothing at all. He insisted that the government should impose only moderate taxes on the emission of CO_2. While this might result in a three-degree (Celsius) increase in average temperatures by the year 2100—a number that climate scientists shudder to imagine—he considered this policy the best approach for optimizing human welfare over time. The key consideration in Nordhaus's model was how to set the discount rate—knowing that, the higher the rate, the less future welfare is valued today. Nordhaus argued that 5.5 percent is a sensible number, while one of his foremost critics, British economist Nicholas Stern, insisted that 1.4 percent is more reasonable. The difference in terms of policy-implications is tremendous. If Nordhaus's number is used, the optimal spending today to offset a million dollars of damages in fifty years is $70,000, whereas if Stern's number is used, the optimal spending increases to a half-million dollars.[74] In short, when economists actually did take the environment into consideration, they did not listen very carefully to what environmental scientists had to say about the likely effects of climate change, and therefore ended up falling short of proposing suitably aggressive measures to restore a more viable balance between nature and economy.

Conclusion

Equipped with general equilibrium theory and growth theory, neoclassical economics showed how people living in a world of scarcity, with the help of science and markets, could optimize the relationship between the economy and nature, and also ensure that the vision of cornucopianism was kept alive. As the twentieth century wore on, neoclassical economics gradually crowded out all other approaches to economics. Aided by the formation of professional organizations, academic journals, university departments, and academic prizes (most notably the Nobel Memorial Prize in Economics), neoclassical economics became near-hegemonic, at least in American and European universities.[75] Crucial to the coronation of economics as the "queen of the social sciences" was MIT economist and, yes, Nobel laureate Paul Samuelson (1915–2009). His enormously popular textbook *Economics* (1948) synthesized the neoclassical approach and made it accessible to students around the world.[76] Postwar neoclassical economics succeeded in pivoting the analytical emphasis of economics from work to consumption, from workers to consumers, from history to statics, from sociology to psychology, from class to individualism, and from narrative to mathematical analysis.

As a result of the dominance of neoclassical economics in the West, anyone who has enrolled in an economics course in the last few decades has been told that scarcity is an unavoidable and universal fact of life.[77] Textbook authors never tired of repeating Lord Lionel Robbins's explanation of Neoclassical Scarcity: "We have been turned out of Paradise. We have neither eternal life nor unlimited means of gratification. Everywhere we turn, if we choose one thing we must relinquish others. Scarcity of means to satisfy given ends is an almost ubiquitous condition of human behavior."[78] Students of economics have been told that Neoclassical Scarcity spontaneously incentivizes people and firms to optimize the use of their resources. Over time, they perpetuated the process of economic growth and thus contributed to the gradual improvement in standards of living across the globe—as long, that is, as human creativity remains robust. We are living through the neoliberal "end of history," a world of abundance forever ruled by Neoclassical Scarcity.[79] In the words of the heterodox Harvard economist Stephen Marglin, growth is constantly chasing scarcity, but "like the mechanical rabbit at the dog races, Scarcity has always managed to stay comfortably ahead of its pursuer."[80]

The neoclassical version of scarcity served as a powerful intellectual inspiration for the emergence of neoliberalism in the 1970s, which soon became a potent political force in the hands of Margaret Thatcher and Ronald Reagan. As Reagan confidently declared, there are "no great limits to growth because there are no limits on the human capacity for intelligence, imagination, and wonder."[81] Neoliberalism continued to build momentum with the demise of the Soviet Union and the embrace of the Washington Consensus in the realm of international politics. By the 1990s, economists sanctioned escalating consumption and economic growth as the legitimate aims of every society. As capitalism extended its hegemony across the world, Neoclassical Scarcity became all the more ideologically powerful.

8

PLANETARY SCARCITY

On Christmas Eve 1968, astronaut Bill Anders photographed the earth as it appeared on the horizon of the moon. Then, Anders and fellow astronauts Frank Borman and Jim Lovell took turns reading the first ten verses of Genesis in a broadcast heard by millions back on earth. They finished with a blessing for the season: "good night, good luck, a Merry Christmas, and God bless all of you—all of you on the good earth."[1] Though the astronauts of Apollo 8 did not read beyond the tenth verse, many in their audience would have been familiar with the rest of the book of Genesis: On the fifth day, God made Man in his image and likeness to exercise dominion over all living things. This bond between the Creator and humanity would persist even after Adam and Eve were expelled from the Garden of Eden and God sent a deluge to drown the world. Carried aloft by a combination of kerosene and liquid oxygen, Anders, Borman, and Lovell were the spacefaring descendants of Adam and Noah, blessed by God to carry Man's dominion from the earth to a new world.

Yet such confidence in Christian cosmology did not entirely shield the astronauts from another, far more unsettling discovery. In the live

Earthrise, **December 24, 1968.** Bill Anders's photograph encapsulated the clash of two opposing ideologies, expressing at the same time a celebration of human power over nature and a growing concern about the fragility of the biosphere. *Credit:* NASA.

broadcast to earth, Jim Lovell spoke of the "awe-inspiring" effect of watching the dance of the celestial bodies. "It makes you realize," he mused, "just what you have back there on Earth." Still thinking of the Old Testament perhaps, he conjured up an image of water and verdure in the desert: "The Earth from here is a grand oasis to the big vastness of space." Reflecting on the same void, Anders felt his faith waver: "We are like ants on a log. . . . How could any earth-centered religious ritual know what God's truth is?"[2] This startled sense of wonder was even more apparent in Anders's photograph of earth rising on the moon's horizon: a tiny blue island in a sea of nothingness. The image revealed the concrete and indissoluble unity of the Earth's biosphere; within its razor-thin atmosphere, blue oceans, white clouds, and glimpses of land appeared, but no political boundaries or social divisions. Over the coming decades, Anders's "Earthrise" photograph would become an icon of environmental consciousness, encapsulating the fragility of life and the need for planetary stewardship. In parallel with this political movement, the new interdisciplinary field of earth system science began to map how the planet functioned as an integrated system. Humans were not lords and masters of creation but utterly dependent on the life support of the biosphere. The view from orbit drove home how little of the earth system humans actually understood, much less controlled.[3]

These conflicting impulses of the Apollo mission—technological dominion and environmental fragility—brought to the surface a deep contradiction of late-twentieth-century society. After the carnage of World War II, the advanced economies of the world entered a period of unprecedented economic growth powered by new technology, cheap energy, and mass consumption. At the same time, human impacts on the environment escalated, setting the planet on the path toward multiple dangerous tipping points. The quickening pace of economic growth in the postwar era was the outcome of several closely connected forces. Competition between the West and the Soviet bloc created a strong incentive to maximize growth through technological change. The American contribution was crucial, of course, but equally important was the parallel development of Soviet industrialization. In both cases, intensifying energy use went hand in hand with economic expansion and new kinds of consumption. Fossil fuels—coal, oil, and natural gas—provided the bulk of the cheap energy required to power an unprecedented pace of urbanization, crowned by the proliferation of megacities after 1950. Fossil fuels were also critical to industrial agriculture, which saw very high inputs of energy for every calorie produced.[4]

Latter-day scholars have christened this phenomenon "the Great Acceleration." In scientific terms, the concept captures the systemic and interrelated impacts of economic development on the biosphere that began around 1950 and are still continuing to the present. It is closely connected to the concept of Planetary Boundaries put forward in 2009 by environmental scientists Johan Rockström and Will Steffen (as we discussed in the Introduction). Vital socioeconomic and environmental indicators show steep growth curves in world population, real GDP, primary

energy use, transportation, water use, and tropical forest loss. A vertigi-
nous rise in greenhouse gases marked the most ominous of these changes.
Atmospheric concentration of carbon dioxide had surpassed the Holocene pattern of natural variability for the first time at the end of the
nineteenth century, but after World War II carbon emissions increased
much more rapidly, from a global mean of 311 ppm in 1950 to 331 in
1975, reaching 417 by 2022. In the first decades of the postwar era, most
emissions came from Europe and North America. After 1980, China,
India, and other developing economies contributed increasing shares.[5]

Politicians and economists greeted the Great Acceleration with open
arms. For most members of the economics profession, the trajectory of
sustained growth between 1950 and 1970 seemed to vindicate the basic optimism of the discipline about the truth and universality of Neoclassical
Scarcity. Apparently, utility-maximizing consumers and profit-maximizing
firms had spontaneously settled on the optimal resource allocation. Innovation, substitution, and growing efficiency seemed capable of averting
the threat of depletion in nonrenewable resources like copper, tin, iron
ore, and petroleum. The new aggregate measure of Gross Domestic Product made abstract growth easy to fathom and celebrate. In liberal de-
mocracies, the holy grail of sustained growth became the foundation of
electoral politics in these giddy decades. Economists achieved newfound
prominence as the guardians of future progress, and students flooded economics departments.[6]

Yet with the Great Acceleration came increasing disquiet among a
vocal minority of social critics and scientists. Eminent philosophers
warned about the corrosive effects of growth-oriented culture on human
values and institutions. Rapid population growth revived Malthusian
worries about the physical limits to the economy. Ecologists grew increasingly concerned about the effects of man-made toxins on the biosphere.
Scientists and computer modelers began to map out the interactions
within the earth system, exploring the circulation of carbon and other
cycles. By the 1980s, the depletion of the ozone layer and the increase of
greenhouse gases in the atmosphere indicated that the biosphere was
much closer to disruption than earlier generations had imagined. In 2000,
Paul Crutzen and his colleague Eugene Stoermer proposed that humanity
had left the relatively stable and benign climate of the Holocene to enter a

new dangerous epoch, which they named the Anthropocene. This presented a visceral challenge to the mental universe of neoclassical economics. Insatiable wants and endless growth were on a collision course with the earth system itself. The world of Neoclassical Scarcity now confronted a rising consciousness of the condition we call Planetary Scarcity.[7]

This turn to Planetary Scarcity was centered on the unsettling discovery that runaway extraction and consumption produced pollution on such a great scale that human waste was overwhelming the cycles that kept the planet habitable and hospitable to complex societies. Even the oceans and the atmosphere were filling up with human contamination. What had once seemed a noble mission to conquer nature now looked increasingly misguided and dangerous, rife with unintended consequences and delusions of grandeur. In contrast with Malthusian fears about the pressure of population on the finite supply of land and mineral stock, Planetary Scarcity brought attention to the burden of overconsumption on the physical sinks of the earth.

This chapter traces the intellectual emergence of the concept of Planetary Scarcity across the era of the Great Acceleration. Already in the late 1950s and early 1960s, critics like Hannah Arendt and Rachel Carson warned that humans were overwhelming the earth with new kinds of nuclear pollution and chemical toxins. Cold War scientists played a critical role in supplying evidence about disturbances in the oceans, the atmosphere, and other natural systems. A new science of carbon exchange was developed for the atmosphere and the ocean from the 1950s onward using a language of sinks and reservoirs. In the 1980s, scientists awakened the public to the planetary threats of greenhouse gas emissions and ozone depletion. The problem of protecting planetary sinks from industrial pollution became a serious political issue. Such worries have only grown deeper in recent decades.

Yet the idea of Planetary Scarcity has also spurred a positive expression—a search for alternative ways of understanding nature and the economy. This movement was underway already at the start of the Great Acceleration. A dawning awareness of ecological risk quickly seeped into postwar social theory. This chapter therefore weaves two histories together: the story of Planetary Scarcity's emergence and the parallel history of postwar critiques of consumer society, including important new movements in ecofeminism, ecological economics, and economic anthropology.

The Philosopher in the Great Acceleration

The Great Acceleration did not go unnoticed by philosophers. One of the earliest sustained treatments of the phenomenon came from the German-American thinker Hannah Arendt (1906–1975). Born in Hanover to a middle-class Jewish family, Arendt led a life that encapsulated the cataclysms of the twentieth century. After completing her studies with Martin Heidegger and Karl Jaspers, she became a stateless refugee from the Nazi regime. She made her name in American academia with her dazzling dissection of fascist politics, *The Origins of Totalitarianism* (1951). Arendt's next major work, *The Human Condition* (1958), mined the tradition of classical political philosophy to deliver a piercing attack on the modern cult of prosperity and progress. The book opened with the launch of Sputnik in the fall of 1957—the first satellite to orbit earth. For Arendt, Sputnik was a world-changing event, even more significant than the splitting of the atom, not because the Soviets had beaten the Americans into space, but because the dream of escaping the bounds of the earth now seemed within reach. Modern science with its godlike powers finally seemed to have fulfilled the Cornucopian promise of Bacon and Hartlib. But this was not an unmitigated blessing. Science had brought into the human realm *cosmic forces* (by which Arendt meant nuclear energy) that threatened to overwhelm the natural and social world. Arendt's observations on technology were more prescient than she could have known at the time. The ascent of Sputnik coincided with the first attempt to map greenhouse gas emissions during the International Geophysical Year (1957–1958). Unbeknownst to Arendt, the scientist Charles Keeling had begun to measure the buildup of carbon dioxide in the earth's atmosphere a few months after Sputnik was launched. Over the next decades, rocket technology made it possible to establish a network of satellites to monitor weather on a planetary scale. Sputnik thus marked the beginning of a global infrastructure of information gathering, which in turn ushered in the emergence of modern climate science.[8]

Although Arendt's political theory was not directly ecological in inspiration, she shared with Rachel Carson and other environmental writers the recognition that the deep structure of human praxis and thought depended on the planet itself. "The earth," Arendt observed, was "the very quintessence of the human condition." Though humans had set themselves apart from animals through the power of artifice, they shared with animals and all other life the same dependence on the support system of the biosphere. Earth provided "human beings with a habitat in which they can move and breathe without effort and without artifice." From the natural order of the planet flowed the basic order of human communities: birth, life, and death.[9] At the same time, Arendt observed that the human endeavor required a separation between the natural and artificial sphere—as she called it, "the time-honored protective dividing line between nature and the human world." Artifice was a form of violence that wrested from the meaningless cycles of nature a durable world of objectivity and meaning. Arendt subdivided the active life of humans on earth into three "basic conditions."[10] *Labor* was the realm of necessity governed by the biological rhythms of the body. *Work* was the domain of the artist and the artisan who created durable things and made the earth into a dwelling fit for human beings. *Action* was the realm of choice—the arena where courageous individuals made history through political contestation.

What worried Arendt most about the state of modern society was the threat that unrestrained consumerism posed to the earthly balance of labor, work, and action. Technoscience and capitalism had elevated labor to the point of being the only worthwhile human activity. The first step in this process occurred when industrial labor replaced craftsmanship during the Industrial Revolution. By the middle of the twentieth century, modern science had achieved such power that it could channel natural forces directly into society. In giving free rein to human "needs and wants," automated production undermined the stability and meaning of the man-made world, replacing it with a consumer order oriented toward endless obsolescence:

> For a society of laborers, the world of machines has become a substitute for the real world, even though this pseudo world cannot fulfill the most important task of the human artifice, which is to offer mortals a dwelling place more permanent and stable than themselves.[11]

The expansion of the modern economy now covered the entire globe, with no apparent limit in sight: "every end is transformed into a means."[12]

By enshrining endless desire and the mastery of nature—the basic tendency of Neoclassical Scarcity—modern society produced a "waste economy, in which things must be almost as quickly devoured and discarded as they have appeared in the world." Labor and consumption formed "ever-recurring cycles" that eroded the durability of the world, depriving humans of a home on earth.[13] At the same time, the triumph of *Animal Laborans* stripped human experience down to "empty processes of reckoning" until the "only contents left in the mind were appetites and desires, the senseless urges of the body."[14] Though the capacities for art and action still persisted in such a society, they now became superfluous and marginal to human experience. The population sank into a state of "automatic functioning" characterized by a "dazed, 'tranquilized,' functional type of behavior."[15] Whenever laborers gained spare time, it was "never spent in anything but consumption, and the more time left to him, the greedier and more craving his appetites." The end result was a world where consumption consumed all things: "no object of the world will be safe from consumption and annihilation through consumption."[16] In this condition of excessive affluence, the "capacity for action" became the "exclusive prerogative of the scientists" who had little interest in the web of human society.[17] To counter these dangerous tendencies, Arendt argued that labor must be subordinated to work and action. Rampant consumerism must give way to a new ethic centered on "building, preserving and caring for a world that can survive us."[18]

Arendt was hardly alone in calling for an alternative to consumer society. Herbert Marcuse's *One-Dimensional Man* (1964) became a counterculture bestseller and a dorm room bible for the New Left. Marcuse (1898–1979), too, was a former student of Heidegger and, like Arendt, a Jewish refugee from Nazi Germany who found sanctuary from oppression in American academia. Haunted by his experience as a citizen of the short-lived Weimar republic, Marcuse saw postwar American affluence through the lens of his European past. Modern technology had created conditions in advanced industrial society under which all needs could be satisfied. But in both the West and the Soviet Bloc, the promise of freedom from want turned out to be a poisoned chalice. Advanced industrial society was "destructive of the free development of human needs and faculties."[19] In both its capitalist and socialist guise, this technocratic regime suppressed true freedom, by seducing the "vast majority of the population" with "a rising standard of living."[20] Marcuse genuinely believed that modern technology could deliver people from want, but he also saw this new affluence as a terrifying tool of mental oppression. Citizens on both sides of the Iron Curtain lived in a state of false freedom. "Social controls" produced a regime of "false needs."[21]

There was more than a passing resemblance between Marcuse's pacifying and narcotic version of creature comforts and Arendt's vision of dazed and tranquilized automatic life in the society of labor. They differed principally in how they imagined a reconstructed order of genuine freedom: Marcuse's aim was to transcend the currently dominant technological rationality and engage in the liberatory process of critical theory, or what he called "negative thinking," while Arendt looked to ancient philosophy for a path to resurrect political action. For Marcuse, the insatiable desires underlying Neoclassical Scarcity were inauthentic products of capital that entrapped consumers in a state of social subjugation. For Arendt, the trap lay in "mass culture"—and its "universal demand for happiness." According to her, "neither the craftsman nor the man of action has ever demanded to be 'happy' or thought that mortal men could be happy."[22] By artificially accelerating and expanding the rhythms of consumption and production, consumer society eradicated the space for political action and art.

Yet another philosophical critique of consumer society emerged in the writings of their teacher Martin Heidegger (1889–1976). Unlike Arendt and Marcuse, Heidegger eschewed questions of freedom and politics in favor a phenomenological and poetic exploration of how modern technology had come to colonize ordinary life. Heidegger's greatest philosophical contribution was his investigation of the pre-theoretical and social basis of human knowledge, elucidated through the method of transcendental hermeneutic phenomenology in his masterpiece *Being and Time* (1927). In later life, he developed a mystical approach to the problem of Being, which revived and revitalized Romantic Scarcity with a strongly conservative bent.[23]

In "The Question Concerning Technology" (1954), Heidegger tried to show that the system of modern technology at the most foundational level constituted an interpretation of reality—what he called the "enframing" of

being. Technology reduced the natural world to something to be measured and manipulated, a "standing-reserve."[24] By contrast, farmers in
traditional societies depended on the soil to deliver the harvest and the
wind to grind the grain in the mill. They relied on natural processes without forcing them. In *Being and Time,* Heidegger had set forth a novel understanding of human involvement in the world through the model of the
craftsman. Instead of following the dualistic conception of the active
human and objectified natural world in Bacon and Descartes, Heidegger
wanted to understand how practices and skills enabled people to dwell in
the *Umwelt*—a term he used to describe the occupations that characterized
common life.

In Heidegger's later works, the legacy of seventeenth-century Cornucopian Scarcity, with its focus on technology as the instrument of infinite growth, came into sharp focus. In modern society, technology had
transformed the entire world into a storehouse of resources with an aim
to extract "the maximum yield at . . . minimum expense."[25] Modern technology isolated the functional dimension of things—their role as natural
resources—while covering up their essence. In the process, technology
took on a momentum of its own, independent of human intentions. The
force of technology altered and reframed the essence of human life so that
instrumental use pervaded all action and perception. For example, the
construction of a hydropower dam on the Rhine reduced the river into a
mere storehouse for energy. Even though the landscape of the river had not
been entirely obliterated, the spirit of technology inevitably degraded and
deformed perceptions of the natural world. In this sense, modern tourism
replicated the attitude of the engineer: the beauty and grandeur of the
Rhine had become nothing more than "an object on call for inspection by
a tour group ordered by the vacation industry."[26]

Another 1954 essay by Heidegger entitled "Building, Dwelling,
Thinking" offered a poetic solution to the problem of modern technology
by reviving the Romantic notion of living a simple life rooted in the earth.
Since time immemorial, humans had sought shelter from predators and
other threats. This search for protection involved care for other living beings, including livestock and food plants. Yet genuine dwelling required
more than simply building houses. For Heidegger the durable structures
erected by humans in the landscape were "things" in the ancient sense of

assemblies. They gathered together the fourfold forces of the world and remained at peace with them: "The fundamental character of dwelling is this sparing and preserving." Heidegger continued, "Real sparing is something positive and takes place when we leave something beforehand in its own nature, when we return it specifically to its being."[27] Heidegger spoke of a bridge joining two sides of a river as a dwelling in so far as it gathered "the earth as landscape around the stream." It "lets the stream run its course and at the same time grants their way to mortals so that they may come and go from shore to shore."[28] Poetry provided the crucial medium by which to recover the world of dwelling. For Heidegger this meant above all the works of Friedrich Hölderlin, a contemporary of William Wordsworth and John Clare. He closed the essay by describing how a historical example of dwelling—a simple farmhouse in the Black Forest—existed in harmony with the earth and sky and made room for generations to "journey through time"—an idea that harkened back to Rousseau's Edenic islands and Alpine villages.[29] Heidegger seems to have wanted to awaken in the public a sense of urgency about living *with* the earth. Our plight, he noted, lies in the problem that "mortals ever search anew for the nature of dwelling, that they must ever learn to dwell."[30] His defense of dwelling perhaps most closely anticipated the so-called Deep Ecology movement, with its rejection of Cartesian dualism in favor of a quasi-mystical subordination of the individual to the community of nature.[31]

Spaceship Economics

The existential and ecological problem of dwelling with the earth was articulated in a different way by Rachel Carson (1907–1964) in her ecological critique of consumer society. A marine biologist by training, Carson entered government service in the US Fish and Wildlife Division while pursuing a parallel career as a naturalist writer. In a trilogy of works about ocean life, she developed a lyrical sensibility of great power. Her interest in marine biology also stirred an early awareness of the effects of synthetic chemicals on human health and wildlife diversity. These concerns centered on the pesticide Dichlorodiphenyltrichloroethane (DDT), which had come into wide use to fight malaria during World War II. Though DDT was greeted as a marvelous breakthrough at the time and its inventor Paul

Muller received the Nobel Prize in Medicine in 1948, American government scientists soon sounded the alarm about the unintended consequences of the pesticide. Carson's critique of pesticide use in *Silent Spring*
(1962) caused a public sensation and made her a figurehead of a new kind
of conservationism.[32]

Silent Spring opened with the image of an ordinary American town,
seemingly "in harmony with its surroundings." Yet underneath the surface of bucolic peace, there were signs of disturbance. Local vegetation
and wildlife began to succumb to "mysterious maladies."[33] Chicken,
sheep and cattle sickened. Unexplained deaths occurred among farming
families. Even the birds fell silent and vanished. Modern chemistry had
inadvertently unleashed a nightmare threat into the midst of everyday
life. At the time, DDT was virtually omnipresent in American households
and agriculture. Just as the second wave of feminism declared the private
political, Carson showed that environmental risk also began at home. Her
main lesson was that the toxins used to kill kitchen bugs or farm pests
could not be contained within a safe zone. The strategy of DDT spraying
mistakenly assumed such poisons would target only specific species. But
in reality, all species shared the same fundamental biology and therefore
a vulnerability to synthetic compounds like DDT. The unintended consequences of pest spraying showed how human ambitions to master the
environment actually endangered the web of life, exposing the hubris of
chemistry and the willful ignorance of economics, which treated nature
as a cost of production, or an externality at best.[34]

DDT was not simply a local risk. Carson helped popularize the word
"environment" to capture how humans could disrupt and overwhelm
natural systems. In June 1963, just months before Carson succumbed to
metastasizing breast cancer, she testified before Congress about the need
for urgent action to contain the threat of pollution. "Contamination of
various kinds" she warned, "has now invaded all of the physical environment that supports us—water, soil, air, and vegetation." Toxic pollution affected not just wildlife but also the "internal environment" of the human
organism, perhaps across multiple generations. Human pollution of different kinds—radiation, household waste, and pesticides—posed an insidi-
ous threat to the integrity of all life.[35] Moralists and physicians had warned
about the dangers of consumption to the health of individuals and society
in past centuries. Carson's book revealed how chemical compounds tied

to middle-class consumption could poison ecosystems across the earth. DDT accumulated through the food chain, striking top predators like birds of prey. Its menace was compounded by the uncertainty and delay involved in its long-term effects.

The scare about DDT was part of a broader set of worries about other ecological risks and toxins, from growth hormones to nuclear dumping.[36] Carson herself drew attention to the dangers of radioactive waste to marine ecosystems in the revised preface to her book *The Sea Around Us* in 1961. The ocean, like the atmosphere, was not an infinite dumping ground for pollution but a finite sink vulnerable to universal contamination. Waste deposited in one location could easily travel on the currents to pollute the whole marine system. Here and in *Silent Spring*, Carson helped pioneer the new environmental discourse we call Planetary Scarcity. Instead of worrying about the physical limits of finite stock, as Malthus had done more than a century before, Carson shifted attention to the problem of finite sinks for pollution, and to the interconnectedness of all life. The chemical menace of DDT and the nuclear contamination of the ocean thus anticipated the environmental horror stories of coming decades—first the threat of Chlorofluorocarbons to the ozone layer and then the danger of greenhouse gases to the climate of the Holocene.[37]

Concerns about the fragility of the environment were not an invention of the 1960s. It is possible to trace a long history of ideas about natural deterioration and catastrophe stretching into the early modern era. Even older is the tradition of conservation and resource stewardship. What did change in the postwar era was the pace and scale of economic and technological change, along with the emergence of new fields of science, including systems ecology and earth system science. These developments in turn spurred novel ways of grasping environmental vulnerability at the global level and over geological time scales and thus enabling an understanding of scarcity in a planetary frame. Carson's book marked the spread of such ideas into wider popular consciousness in the United States and beyond. However, her message did not always meet with a hospitable reception. Advocates for the chemical industry unsurprisingly defended their turf, often resorting to personal attacks to denigrate Carson's stature. Likewise, the great mass of professional economists also resisted Carson's pessimistic interpretation of technology, for all the reasons explored in the previous chapter. Only a tiny minority of prominent economists took the warnings about environmental degradation as a genuine challenge to Neoclassical Scarcity. While they found themselves marginalized in their own discipline, their approach opened the door to a radical reorientation of economic analysis toward the biophysical context of production and exchange, which coalesced into the field of ecological economics in the 1980s.[38]

One of the first postwar economists to embrace ecological thinking was the British-American academic Kenneth Boulding (1910–1993). After receiving first-rate training in economics at Oxford and Chicago, Boulding seemed destined to climb to the top of the profession. In 1949, he won the John Clark Bates medal, one of the most coveted honors for young economists. Paul Samuelson had been the previous Bates winner and Milton Friedman would be the next. Yet Boulding soon strayed from the fold by insisting on a holistic approach that took in both the social and the natural sciences. By the middle of the 1960s, he had begun to incorporate ecological frames into his economic theory. Boulding also dared to question the primacy of mathematical modeling at the moment that it gained ascendancy. For Boulding, this meant jettisoning the basic assumptions of modern economics in favor of an evolutionary and nonequilibrium model. Economics should not imitate the celestial mechanics of the static solar system. Instead, economists should look to the "profound indeterminacy" of evolutionary biology to understand the critical place of the environment in economic change.[39] Boulding cast Adam Smith as the founder of evolutionary economics (perhaps not such an outrageous idea when we remember the centrality of agriculture in Smith's model of growth).

Thermodynamics inspired Boulding's classic challenge to Neoclassical Scarcity, "The Economics of the Coming Spaceship Earth" (1966). Here, Boulding contrasted two ways of thinking about scarcity: the open "cowboy economy" of the past and the closed "spaceship economy" of the future.[40] The roots of the "cowboy economy" went deep. Perhaps inspired by Christian theology, Boulding attributed it to a universal and permanent tendency in human nature. Yet he also left little doubt that modern science and technology hade greatly strengthened the impulse: "The extraordinary achievement of the last 200 years have given us certain delusions of grandeur and a certain feeling that man can accomplish anything if he only puts his mind to it." Communists took this belief to the greatest extreme with their faith in human "infallibility, omnipotence, and immortality."[41]

But they were far from alone. In the West, economists, politicians, and ordinary consumers happily rejected all notions of limits. "Economists, and indeed mankind generally, have tended to treat the economic system as if it could enter into continuous exchange with an infinite reservoir of nature." Yet this cornucopian belief was now growing untenable. Ecology had brought these "flights of omnipotent fancy . . . down to earth."[42]

Boulding imagined the closed system of the spaceship model in terms of a cyclical economy where all materials were recovered (though energy remained subject to entropy). The aim was to minimize throughput while maintaining stock. There were "no mines and no sewers" in a spaceship.[43] In an astonishing reversal of basic economic dogma, Boulding proposed that the processes of "production and consumption" on the whole must be counted as "bad things" rather than essential measures of success in the economy.[44] In the spaceship economy, "consumption is no longer a virtue but a vice, and a mounting GNP is to be regarded with horror."[45] For Arendt, the event of space travel seemed to promise transcendence from earthly conditions, but for Boulding, the actual practice of space travel was necessarily an ecological problem. A voyage between the stars involved the task of preserving life within a closed space over many generations—maintaining a circular economy by drawing on a "very small stock" that "circulated constantly through the system."[46]

The turn toward spaceship economics also brought the rights and claims of future generations into focus. Boulding saw pollution as a pressing problem rather than an issue that could safely be adjourned for future engineers to resolve or internalize into the economist's cost function. He presciently noted that pollution in the atmosphere might become a "major problem in another generation." Indeed, "even today it is clear that oceans and the atmosphere are by no means inexhaustible reservoirs."[47] In parallel with Hannah Arendt and Rachel Carson, Boulding was articulating a new idea of Planetary Scarcity. With eerie accuracy, Boulding suggested that it would be "fatally easy" for people to "change the composition" of the ocean or the atmosphere "in such a way that the earth will pass some watershed point—for instance, through something like the greenhouse effect of the accumulating carbon dioxide in the atmosphere— which will destroy the existing equilibrium which may be much less desirable for man."[48] These tipping points were fundamentally connected with demographic pressure as well as growing consumption. A skyrocketing world population increased the "chance of man's activities seriously in-terfering with the whole balance of the planet."[49] Before the Industrial Revolution, when only a few hundred million humans inhabited the earth, there had been space for most other species to thrive, but the rapid accelera-tion of mankind in the twentieth century made this balance increasingly precarious. The world population had increased from two billion in 1930 to three billion in 1960, and would reach four billion in 1975. Boulding predicted that the "insatiable pressure for food supplies" would cause a mass extinction of animals and plants in the next century.[50]

From the perspective of the "spaceship economy," the concept of scarcity took on a radically new meaning. Rejecting the fungible world of neoclassical economics, with its infinite substitutability, Boulding saw the economy as a subset of a planetary system defined by biophysi-cal feedback and evolutionary change. He replaced the present-centric time scale of Neoclassical Scarcity with a cosmic outlook defined by the forces of evolutionary and geological change. The "technical achieve-ments of the last 200 years," Boulding noted, had largely been made possible through the depletion of "geological capital."[51] This gigantic consumption of iron ore and fossil fuels was a one-time event, never to be repeated (on a human time scale). The question was how best to take advantage of such a gift while moving toward a circular economy after the end of fossil fuel. Paradoxically, the future was at the same time ascetic and cornucopian, Boulding seemed to think. Welfare would come to de-pend not on the speed and scale of throughput but on "the richness and variety of . . . capital stock, including of course . . . human capital."[52] When conservation and durability became primary duties and virtues, a new kind of affluent society would emerge:

A space ship society does not preclude . . . a certain affluence, in the sense that man will be able to maintain a physical state and environment which will involve good health, creative ac-tivity, beautiful surroundings, love and joy, art, the pursuit of the life of the spirit, and so on. This affluence, however, will have to be combined with a curious parsimony. Far from scarcity disappearing, it will be the most dominant aspect of the society. Every grain of sand will have to be treasured, and the waste and profligacy of our own day will seem so horrible that our descendants will hardly be able to bear to think about us, for we will appear as monsters in their eyes.[53]

Because Boulding had so little to say about the psychological drives and rewards of consumption, it is difficult to tell precisely what he meant by affluence and parsimony in relation to human welfare. What we do know is that he resisted crude generalizations about self-interest and material gratification. In neoclassical economics, the ideal subject engaged in self-indulgent processes of utility maximization, in which there were no moral constraints on their hedonism. By contrast, Boulding insisted on the human capacity for selfless sacrifice and savaged utilitarian ideas of preference. Note his gleeful use of Wordsworth's quip: "High Heaven rejects the Lore of Nicely calculated Less or More."[54] As a practicing Quaker, Boulding believed that economic behavior was embedded in a larger "integrative system" extending to values such as "status, respect, love, honor, community, identity, legitimacy, and so on."[55] Without this cultural envelope, economic relations could never have evolved in the first place. Even the most basic form of exchange demanded "trust and credibility."[56] Such norms defined the boundaries of markets as well by demarcating what could be bought and sold. In the "cowboy economy," commodification would proceed endlessly outward. In the "spaceship economy," there was a limit beyond which markets could not extend.

Resource Panics and Counterculture

The photograph "Earthrise" became an icon of earthly fragility at a moment when Malthusian Scarcity once again was gaining force. In early 1968, two American biologists—Paul and Anne Ehrlich—published a chart-topping prediction of famine entitled *The Population Bomb*. The book was written in part to influence the presidential election that year by warning the public about the danger of demographic overshoot. Echoing the mathematical confidence of Malthus, the Ehrlichs painted a terrifying picture of near future calamity, insisting that the "battle to feed all of humanity is over." In the next decade, they argued, the world would undergo massive famines, which would kill hundreds of millions of people. For the Ehrlichs, any real solution to the problem required population control: "The first move must be to convince everybody to think of the earth as a space ship that can carry only so much cargo." When their prophecy proved false, the fiasco gave a boost to counterarguments about the dangers of Malthusian pessimism. Critics pointed to the success of the Green Revolution in increasing agricultural productivity across the Global South. Even so, Ehrlich's Neo-Malthusian position remained a potent influence on the environmental movement.[57]

The year of Apollo 8 also saw the founding of the Club of Rome, an international network devoted to investigating pressing contemporary problems. At the suggestion of the American computer engineer Jay Wright Forrester, the group decided to focus on growth as the underlying cause of global crisis. In 1972, the Club of Rome issued its first report based on groundbreaking computer simulations of future trends in population growth, resource consumption, and pollution. Like Paul and Anne Ehrlich, they predicted overshoot in the near future. The Club of Rome's report, *The Limits to Growth* (1972), popularized a systems approach to environmental problems, stressing multiple variables and conflicting rates of exponential change.[58] It made use of scenarios to model different futures, including a world of overshoot as well as a future of ecological equilibrium. When US oil prices rose 400 percent in response to the OPEC embargo of October 1973, the crisis seemed to deliver grim confirmation of the precarious nature of modern growth. Though the cause of the supply shock was political rather than a matter of material exhaustion, policy responses to the oil crisis aligned with broader narratives about environmental risk and the need for a transition to renewables.[59]

One of the most radical reactions to the age of "Earthrise" came from feminist thinkers. They argued that the instrumental approach to the environment underpinning Neoclassical Scarcity and postwar capitalism was a disastrous mistake. In her landmark book *The Death of Nature: Women, Ecology and the Scientific Revolution* (1980), environmental historian Carolyn Merchant (1936–) traced the source of this error back to the mechanization of science in the seventeenth century. Trained as an early modern historian of science at the University of Wisconsin Madison in the 1960s, Merchant found crucial inspiration in second-wave feminism, counterculture social protests, and the environmentalist writings of Rachel Carson and Paul Ehrlich. At the center, *The Death of Nature* was a startling reinterpretation of the origins of Cornucopianism. By rejecting the idea of nature as a nurturing mother and replacing it with a machine, philosophers like Francis Bacon and René Descartes had invented a new ideology of mastery that rested on a false separation between mind and body, male and female, human society and nonhuman environment.[60]

Along with other ecofeminists like Susan Griffin and Val Plumwood, Merchant demonstrated that this kind of dualistic thinking extended into the gendered spheres of work and household. Only male labor was valued in the capitalist economy, while the creative sphere of reproduction in the household was ignored and relegated to insignificance. The formal analysis of market exchange and marginal utility ignored the profound reliance of capitalist production on unpaid work in the household. It also turned a blind eye to the dependence of human enterprise on the ecological productivity of the natural world. In her later writings, Merchant articulated a new ethic of care and partnership with nature. Nature was not a machine to be mastered but an active partner in the web of life. Humans and nonhuman communities must be treated with equal moral consideration in "their mutual living interdependence." The new partnership ethic simultaneously fulfilled "humanity's vital needs and nature's needs by restraining human hubris."[61]

The same moment that gave rise to ecofeminism also saw major new work on the psychological dimension of scarcity. For a long time, mainstream economists had bracketed the problem of mental states in favor of the black box of formal preferences. What mattered was not how consumers actually felt but that their preferences were revealed in consumer behavior.[62] In the early 1970s, this formal understanding of scarcity came under fire on multiple fronts. Major critiques included Marshall Sahlins's *Stone Age Economics* (1972), Richard Easterlin's comparative study "Does Economic Growth Improve the Human Lot?" (1974), and E. F. Schumacher's *Small is Beautiful* (1973). Although vastly different in their scope and disciplinary aim, these critics all addressed an essential question left unanswered by the economists: What defines the true nature of satisfaction?

Stone Age Economics doubled as an investigation of hunter-gatherer society and an ecological critique of neoclassical economics. In the first chapter of the book, Sahlins (1930–2021) took aim directly at Lionel Robbins's idea of scarcity and his assumption of an insurmountable tension between "unlimited wants" and "insufficient means." Economic anthropology revealed an alternative "Zen road to affluence where human material wants were finite and few." In such societies, quite simple technology paired with ecological knowledge sufficed to provide adequate sustenance for all. The possibility of "affluent society" was as old as humanity.[63]

At first glance, Sahlins's critique seemed to rehearse ideas launched two hundred years earlier in the pages of *The Discourse on Inequality*. Rousseau had argued that natural man had no knowledge of the future and therefore no reason to yearn for things; the condition of the household in the primitive state was insular and self-sufficient. While Sahlins's optimistic view of Paleolithic foragers bore a certain resemblance to Rousseau's conception of scarcity, it relied on a new language of quantitative social science that measured nutritional demands and hours of work. For Rousseau, "savage" man was a figure lost in time whose existence could only be imagined through philosophical conjecture. By contrast, Sahlins argued that Paleolithic people could be studied in the flesh in the present. He built his case in part on the quantitative ethnographic fieldwork of Richard Lee amongst the !Kung people of the Kalahari desert in southern Africa. What this research showed was that nomadic foragers lived in a state of relative material plenty. Other investigations of Aboriginal people foraging in Arnhem Land in Australia's Northern Territory demonstrated that adults worked on average no more than five hours per day to collect and prepare food. Their labor provided more than adequate sustenance for the group. In cultural terms, hunter-gatherers put a premium on leisure over additional consumption, making time for relaxing, visiting, entertaining, and other pursuits. Sahlins's analysis rested on an ecological basis. Only by knowing the land intimately could foragers sustain their way of life.

Paleolithic "affluence" admittedly came at a certain cost. Effective foraging required constant movement. Possessions had to be kept a minimum. More disturbingly, people unable to move were left behind. Infanticide and senicide kept population size low. Diminishing returns in foraging made "Malthusian practices" necessary.[64] Yet whatever the price of mobility, Sahlins insisted that it held privation at bay. Lest readers dismiss the Paleolithic system as a special case, Sahlins enlisted hunter-gatherers on the side of substantivist theory against market-oriented explanations. The activities of the !Kung should not be judged by the standards of utility-maximizing consumers and profit-maximizing firms. Like his teacher Karl Polanyi, Sahlins believed that economic activity outside market societies was best understood as a provisioning system embedded in social institutions and normative discourses. Anthropology, archeology, history, and ecology revealed a great tapestry of alternative social systems outside the narrow path of Western capitalism. If anything, the mystery to be explained was not the rationality of Paleolithic foraging but the widespread acceptance of Neoclassical Scarcity. Sahlins ended the chapter on the original affluent society with a satirical swipe at modern economics: "it was not until culture neared the height of its material achievements that it erected a shrine to the Unattainable: Infinite Needs."[65]

Another penetrating criticism of Neoclassical Scarcity came from within the profession. American economist and demographer Richard Easterlin (1926–) brought together survey data from nineteen countries, split between industrialized nations and countries in the developing world, along with a detailed national time series of attitudes about happiness in the United States between 1946 and 1970. Easterlin's findings seemed to contradict a basic assumption of economic progress that a constant rise in GDP should yield a commensurate rise in life satisfaction. Despite significant material improvements in living standards, consumers in the United States reported little subjective change. Subsequent scholarship described a threshold effect for happiness: evidence of subjective happiness increased up to a certain point but then stagnated and remained stubbornly flat despite respectable GDP growth. While Easterlin did not frame his research in an ecological context, these findings found a receptive audience among environmental critics looking to rethink the aims of growth in the face of planetary crisis. Why should economies go on producing more goods and thus destroy the environment when additional consumption could not make affluent people any happier?[66]

In the same moment, German-British economist Ernst Schumacher (1911–1977) became an unlikely prophet of Green counterculture with his wildly popular book *Small is Beautiful: Economics as if People Mattered*, first published in English in 1973 and translated into fifteen languages afterwards. Schumacher was a refugee from Nazi Germany who had served as an economic planner of German reconstruction before becoming a statistician for the British National Coal Board. Side by side with these professional commitments, Schumacher grew interested in questions facing underdeveloped nations, guided in no small part by his spiritual devotion to Buddhism. Visits to Burma and India made a deep impression on

Schumacher. In *Small is Beautiful,* he gathered these experiences into an alternative theory of development that he called Buddhist economics.[67]

Where Western economists put the material standard of living at the center of economic life—ever more goods at lower prices—Buddhist economics instead focused on the twin goals of meaningful work and liberation from desire. Since consumption provided "merely the means to human well-being," the aim "should be to obtain the maximum of well-being with the minimum of consumption."[68] An economy devoted to the endless growth of wants went against the most basic dictum of wisdom: "simplicity and nonviolence."[69] The impetus for Buddhist economics came in part from Schumacher's interest in the problem of securing local livelihood at the level of the Indian village economy. What kinds of investment, technologies, and energy sources were most appropriate to support such communities? The other influence on Buddhist economics came from ecological, romantic, and anarchist thought. Like Mill and Ruskin before him, Schumacher endorsed growth "towards a limited objective" but rejected "unlimited, generalized growth" for its own sake.[70] The appeal to Buddhism tapped into a wider fashion for eastern spirituality in Western counterculture though Schumacher himself had actually converted to the Roman Catholic faith in 1971. His own stance was explicitly ecumenical and pragmatic: all the great spiritual traditions of the East and West, he intimated, served the same purpose as safeguards against the nihilism and folly of materialist economics.

Global Environmentalism

By the early 1970s environmental concerns garnered increasing support around the world. Millions gathered in the United States for the first Earth Day event in April 1970. New organizations such a Greenpeace and Friends of the Earth attracted attention with innovative protest tactics. Environmentalist priorities also began to shape national and international politics. Green Parties emerged across the West, first at the local level and then on the national scene. In the United States, the Nixon administration established the Environmental Protection Agency in 1970. One of its first achievements was to ban DDT in 1972, ten years after *Silent Spring*. Environmentalist politics was not just a concern of the Western middle class. In the Himalayas, local villagers united into the Chipko movement to defend use rights and forest preserves. Over the following decades, environmentalist values went from fringe concerns to mainstream priorities in public opinion and governance across the globe. Yet this pattern at the same time produced an apparent contradiction. The success of counterculture values and the spread of environmental consciousness did very little to dent the basic trends of the Great Acceleration. Indeed, the condition of the earth system began to show serious strain *after* the globalization of environmentalism.

The simple explanation for this apparent contradiction is that environmentalist values and policies often served as excuses to justify continued consumerism. After all, the most common slogan of sustainable development explicitly set out to reconcile economic growth with environmental stewardship. The concept became fashionable thanks to the 1987 UN report *Our Common Future* chaired by Norwegian prime minister and leader of the Labor party Gro Harlem Brundtland (1939–). At the heart of the text was a multigenerational vision of equitable growth: "Humanity has the ability to make development sustainable to ensure that it meets the needs of the present without compromising the ability of future generations to meet their own needs."[71] Essential human needs included "food, clothing, shelter, jobs" as well as an aspiration "for an improved quality of life."[72] Beyond the most elementary definition, the authors made little effort to establish particular quantitative or qualitative parameters of needs and well-being. Brundtland's report instead focused on the long-term challenges to global and intergenerational equity, including deforestation, desertification, fossil fuel exhaustion, climate change, population pressure, and decline in biodiversity. Yet despite this catalogue of risks, the UN commission suggested that economic growth and environmental stewardship remained compatible goals. Brundtland and her colleagues proposed that rising standards of living could be achieved without serious ecological strain.

That sanguine outlook was very much in evidence in a talk on energy policy Brundtland gave at Harvard University in the fall the same year. The North Sea oil supply, she told her audience, represented a precious intergenerational heritage to be husbanded wisely for the long term. Sustainable stewardship entailed a "moderate depletion policy" informed by "mature behavior" with a thirty-year horizon.[73] Brundtland ended her talk by commenting on the work of the 1987 Commission:

The Commission is sounding an alarm, but it does not paint a
gloomy picture of the future. Quite to the contrary: we believe
that human resources and ingenuity, our capacity to address
the issues in a responsible concerted manner, have never been
greater and that we can indeed solve both energy and environmental problems in a new era of economic growth—an era
in which economy and ecology are merged at all levels of
decision-making and where there is a more equitable distribution of wealth within and among nations.[74]

Brundtland's celebration of ingenuity and fossil fuel had much in common
with the anti-Malthusian critique of environmentalism fashionable among
economists in the United States during the Reagan years. It was a direct
rebuke to the melancholy predictions of Paul and Anne Ehrlich and the
Club of Rome. Instead of an era of mass famine or mineral exhaustion,
Brundtland cast the coming environmental crisis as an opportunity for
universal growth.

Conclusion

A year later, on a blisteringly hot day in late June of 1988, James Hansen,
the physicist and director of the NASA Goddard Institute for Space Studies, appeared before the United States Senate Committee on Energy and
Natural Resources. He testified that the earth was currently warmer than
it had been at any time across a century of measurement. The past twenty-
five years had seen the highest temperatures on record, and the four
warmest years had all occurred in the 1980s. In all likelihood, Hansen explained, such a disturbing trend would continue into the future. This
1988 Congressional testimony marked a milestone in the history of climate
awareness. While the science of the greenhouse effect reached back to the
nineteenth century, and continuous collection of carbon dioxide data had
started with Charles Keeling's measurements at Mauna Loa in 1958, it was
only in the 1980s that natural scientists began to put together a definitive
picture of the climate system and its sensitivity to greenhouse gases. At a
conference in Villach in 1985, a general consensus crystallized that carbon dioxide levels might double by the middle of the twenty-first century.
The climate system of the Holocene no longer appeared to be stable. What

Hansen did was to raise public awareness of these scientific apprehensions and to turn the topic of global warming into a political issue during the 1988 American presidential election. The same year also saw the establishment of the Intergovernmental Panel on Climate Change (IPCC), which soon became the most important institutional force in building a scientific consensus about the causes and effects of global warming.[75]

In the 1990s and early 2000s, scientists and activists began to experiment with new modes of representation to make climate change and other kinds of planetary degradation visible and compelling to the public. Mathis Wackernagel and William Rees pioneered the concept of the ecological footprint in 1996. This accounting tool highlighted how much land or biocapacity was needed to sustain a specific level of consumption. In 2000, Paul Crutzen and Eugene Stoermer coined the term *Anthropocene* to bring home the dramatic scale of change in the earth system. Around 2005–2006, a variety of experts and institutions began to promote the carbon footprint formula to measure the carbon dioxide emissions caused by particular economic activities. Parallel to this work, the British Department for Environment, Food, and Rural Affairs developed the concept of the social cost of carbon. This formula estimated the net present value of the damage of one additional ton of carbon to the world over the next century. At the same time, a research network led by environmental scientist Johan Rockström and Will Steffen incorporated the Anthropocene idea into a quantitative model of biophysical limits to development. Called the Planetary Boundaries framework, the model described nine major tipping points that would take earth out of its Holocene state (see Figure I.1). Climate change was only one of many possible disruptions to the safe functioning of the earth system.[76]

These new approaches sought to visualize and quantify the planetary impact of the Great Acceleration by tracking its ecological consequences across different scales. Each model encouraged a novel scalar imagination, which situated the individual and the economy inside the biogeochemical processes that maintained the earth system. Each model also involved prescriptions for slowing or limiting dangerous forms of growth—by reorienting energy use, for example, or reinforcing protections to preserve biodiversity. Yet the deeper our knowledge of the earth system, the more grave the challenge appeared. Because carbon emissions permeated the entire economy, the need for constraints and limits necessarily extended to every kind of economic activity.

The political discovery of anthropogenic climate change pushed the discourse of natural limits and environmental degradation in a new direction. From the age of Malthus onward, pessimists had worried about the physical limits to growth imposed by the finite supply of land and nonrenewable mineral stock—the condition we have called Malthusian Scarcity. But with global warming, the threat shifted decisively from a problem of finite stock to a dearth of sinks. Simply put, there was too much coal, petroleum, and natural gas in relation to the earth system's capacity to absorb the waste products of fossil fuel. The ancient symbol of infinite space—the boundless ocean and endless atmosphere—turned out to be all too finite. Rachel Carson and Kenneth Boulding had anticipated the idea of Planetary Scarcity with their warnings about oceanic and atmospheric pollution in the 1960s; Jim Hansen and the IPCC showed that planetary sinks were already in the process of filling up. The discovery of anthropogenic climate change also sharpened the rift between human history and evolutionary time that Rachel Carson had explored in *Silent Spring*. For Carson, pesticides imperiled a biological system calibrated to adapt on an evolutionary time scale. With climate change, fossil-burning humans acted as a geological force in the earth system. Carbon emissions from fossil-fuel growth threatened to disrupt the planetary carbon cycle and the relative stability of the Holocene climate.

Climate change involved social and spatial disjunctures as well as temporal lags. Developing countries were more vulnerable to climate change than affluent nations. Class, gender, and race played a role in shaping the geographies of risk. In addition to such spatial inequalities, the delayed effects of climate change created a temporal divide, as future generations would suffer the consequences of consumption patterns in the present. Carbon emissions eluded easy management since they were the product of myriad interrelated processes in the fossil fuel economy, including not just industrial production and transport but also agriculture, construction, and energy-intensive services. The planetary scale of the phenomenon transcended conventional forms of environmental activism. Even as certainty grew about environmental change and social impacts, powerful political headwinds thwarted effective action, reducing the chance of successful mitigation.

Despite the warnings of the climate scientists, mitigation policy fell far behind the carbon curve. Economists and politicians across the fossil-fuel economies tended to downplay the risks of climate change and the urgency to act. Some even went so far as to deny the reality of global warming. Meanwhile, carbon levels continued to climb upward. An increasing portion of these emissions came from the developing countries, including the new manufacturing powerhouses of China and India. In his 2016 essay *The Great Derangement,* the Indian novelist and social critic Amitav Ghosh (1956–) observes that the rise of the industrial economies in the Global South exposes the cruel truth about fossil-fuel development. By driving up worldwide demand for energy and resources while at the same time increasing the amount of waste and pollution in the system, globalization pushes the planet ever closer to calamitous degradation.[77]

Ghosh's paradox (the promise of global economic growth will produce planetary catastrophe) captures something essential about the relation between Neoclassical and Planetary Scarcity in the Great Acceleration. The more that developing nations seek to emulate the fossil path of the affluent countries, the greater the disruption of the carbon cycle. As emissions have mounted, it has become evident that the promise of the Great Acceleration cannot be universalized without deepening danger to the biosphere. Where Neoclassical Scarcity sees history as the confluence of insatiable, ever-expanding desires and technological progress, Planetary Scarcity reveals the limits of human ingenuity, the power of unintended consequences, and the fragility of all earthly things.

CONCLUSION

Toward an Age of Repair?

In the upper layers of earth's oceans, single-cell organisms collectively known as phytoplankton thrive in great numbers, drifting along the currents. Although microscopic in size, they add up to a biological force that shapes the conditions of life on the planet to an extraordinary degree. Decomposing phytoplankton yield a crucial ingredient in the formation of shale and petroleum over geological time and therefore a major portion of fossil fuel consumption in the present. As converters of solar energy into organic materials, phytoplankton also have a central place in the food chains of the sea. By transforming carbon dioxide and water into oxygen and carbohydrates through photosynthesis, phytoplankton offer a sink for human pollution while at the same time making the atmosphere breathable to other life forms. They absorb about forty-five to fifty gigatons of inorganic carbon each year and produce half of the oxygen in the atmosphere. Nearly 40 percent of carbon dioxide emissions from humans have been taken up by the ocean. Even though they comprise less than 1 percent of the biomass on the planet, phytoplankton are as important to the earth system and the carbon cycle as the rain forests and other terrestrial sinks.[1]

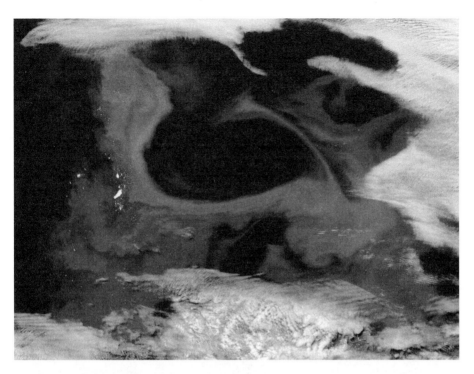

Phytoplankton bloom in the Southern Ocean. *Credit:* Lawrence Berkeley National
Science Laboratory.

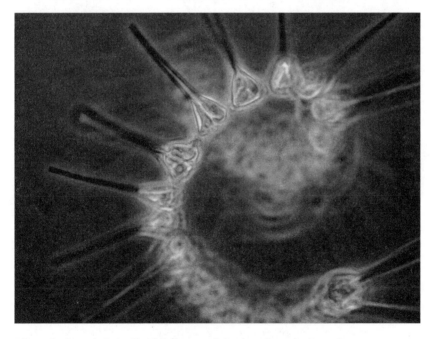

Phytoplankton (microscopic). Microscopic in size, phytoplankton play
an extraordinary role in the maintenance of the earth system. *Credit:* NOAA MESA
project, 1973.

The science of phytoplankton is a belated achievement. Only recently have scientists begun to grasp the significance of these organisms in the earth system. Such knowledge has already sparked hopes of harnessing the species for economic purposes. Some scientists want to make renewable fuel out of phytoplankton genes, by mimicking and accelerating natural processes. Others want to fertilize phytoplankton by dumping iron into the seas to increase the ocean's capacity to act as a sink. However, these ambitions to engineer phytoplankton and incorporate them into the capitalist economy fail to reckon fully with the fragility and complexity of ocean life. Iron fertilization might trigger ecological perturbations of a harmful kind. At the same time, climate change threatens marine biodiversity. Warmer surface waters contain fewer nutrients for phytoplankton, slowing down its growth. Warming waters also fail to mix with colder layers of the ocean, turning off the pump that sequesters carbon in the depths. What will happen to the carbon cycle and the oxygen supply if phytoplankton drastically shrink in numbers? It is probably better if we never have to face this possibility in reality.[2]

For millennia, humans have ignored the depths of the oceans, imagining that the sea was infinite and impervious to human influence. Now we are beginning to appreciate just how vulnerable marine ecosystems are and how much they do to keep the world habitable. The capitalist economy has brought about rapid and massive changes that threaten to overwhelm the earth system. The science of phytoplankton resembles in this regard the discovery of anthropogenic climate change. Both reveal critical material boundaries for human flourishing. Both suggest how little we know about the natural world and how dangerous the current economic ideology has been for the stable functioning of the earth system. Rather than press ahead with further exploitation, a better strategy would be to repair the damage done and then back off to preserve marine ecosystems from further disturbances.

This need to repair becomes even more urgent once we recognize that the threat to the earth system extends to many more domains beyond marine life and carbon emissions. In the Planetary Boundaries model, environmental scientists warn about irreversible and nonlinear changes to the earth system in nine areas: land system change, biodiversity loss, climate change, oceanic acidification, the supply of freshwater, aerosol loading, ozone depletion, nitrogen/phosphorus, and "novel entities." Planetary

Boundaries represent approximate quantitative values for thresholds of environmental risks beyond which we can expect irrevocable change on a continental or global level.

Most prominently, climate scientists have warned that atmospheric CO_2 concentrations above 350 parts per million (ppm) take us out of "the safe operating space" provided by a "Holocene-like state."[3] Beyond the 350 boundary, drastic changes await: glaciers and ice caps will melt, sea levels rise, forest and brush fires become more extensive and destructive, and hurricanes gather force from warmer oceans. If the average temperature rises above two degrees Celsius, coastal cities around the world, including New York City, Miami, Venice, Stockholm, Tokyo, Mumbai, and Hong Kong, might face a sea level rise of between 1 and 2 meters by 2100. Traditional food growing areas would risk losing their capacity to sustain large harvests, triggering subsistence crises in the hottest part of the world, including South Asia and Sub-Saharan Africa. Scientists expect mass mortality events as combined heat and humidity reach wet bulb temperature of 35°C, which is beyond the physiological limit of human endurance. Heat, water, fire, and dearth might put in motion a mass-migration of destitute people, the size of which will make the migration sparked by the civil war in Syria look trivial. The UN predicts that there will be 200 million climate refugees by 2050. This will test the capacity of the political and social systems of the Global North in ways that seem likely to intensify xenophobia and racism, judging by past experience.[4]

To confront these emerging threats, the idea of Planetary Scarcity invites us to reorient and reimagine the purpose of the economy by embracing caution and constraint. Our confidence in mastering the environment has relied all along on a radically incomplete understanding of the natural world. Earth system science undermines this self-assurance by demonstrating how the ideology of maximum efficiency, infinite substitutability, and infinite growth threaten the very processes that keep the planet habitable. This discovery alters our sense of the past as well as the future. The Industrial Revolution produced environmental risks that have only become fully known after more than two centuries of growth. The marginalist economists simply assumed that industrialization could be carried out without dangerous consequences to the natural world, yet unbeknownst to them, carbon emissions from the new global economy had already departed from the pattern of Holocene variability by the last quarter of the nineteenth century. This divergence of neoclassical economics from physical reality only increased in the twentieth century. Lionel Robbins and Paul Samuelson produced their canonical definition of the idea of Neoclassical Scarcity right at the outset of the Great Acceleration. As such, we might think of the persistence of Neoclassical Scarcity as a relic of the Holocene epoch. In the midst of the ongoing rupture of the earth system, neoclassical economics clings to an ideology of human mastery increasingly out of tune with the predicament of the planet. The Industrial Revolution was not a conclusive triumph over nature, but a temporary reprieve bought with fossil fuel energy and a Pandora's box of unintended consequences.

One lesson of the science of greenhouse emissions is that we might not know the extent of future dangers that could be unleashed by present technologies until it is almost too late. What if our technical fixes produce problems even greater than the ones they seek to solve?[5] At this moment in time, one option is to continue embracing Cornucopian optimism, which has been a dominant force in Western economic thought for the last three hundred years. Yet a countermovement is also gaining force, fueled by alarming signs of earth system change. People across the globe are realizing that radical change is needed. Will the new generation refuse the theories handed down to them and instead commit to the formulation of Finitarian ideas, policies, and praxis?

The Holocene Hangover

The need for a novel way to think about the relationship between nature and the economy has yet to make much of an impact on mainstream economists who "are inclined to believe" that market forces "will go a long way toward solving any environmental problems."[6] In the case of resources, such as oil or rain forests, becoming scarce, economists predict that the resulting price increase will lower demand for the goods produced with these natural resources as inputs and will spark greater investments in research and development that will ultimately yield substitutes. They believe that by reducing demand and incentivizing the development of alternatives, the market dynamic, when combined with scientific and technological development, has the capacity to resolve the problem of resource exhaustion. While often optimistic about this dynamic, economists admit

that it will take some time, but that there is plenty of historical evidence
suggesting that the market will be able to work its magic over and over
again. In their bestseller *Abundance: The Future is Better than You Think,*
Peter Diamandis and Steven Kotler argue that humanity lacks sufficient
patience and optimism.[7] Indeed, they suggest that people suffer from what
the Nobel laureate economist Daniel Kahneman calls an "anchoring problem." Because humans extrapolate and linearly project on the basis of
their immediate experiences, they tend to be anchored in the present, failing to imagine future solutions. This "negativity bias," Diamandis and
Kotler insist, makes people overly nervous about ominous prognostications such as those made by Paul Ehrlich in *The Population Bomb* and the
Club of Rome in *The Limits to Growth.* While Diamandis and Kotler do acknowledge remaining challenges, they refuse to believe these cannot be
handled by the miracle of the market and the wonders of science.[8] According to them, accelerating scientific breakthroughs in "computational
systems, networks and sensors, artificial intelligence, robotics, biotechnol-
ogy, bioinformatics, 3-D printing, nanotechnology, human-machine inter-
faces, and biomedical engineering" have the capacity to create a world in
which "the vast majority of humanity" will be in a position to "experience
what only the affluent have access to today," and to do so without destroying the environment.[9] Scarcity is thus the mother of invention; what was
once scarce will become abundant in the future. This fervent belief in
markets and science recalls the exuberant visions of progress dreamt
up by the seventeenth-century alchemists.

The problem with the consumption of fossil fuels, according to modern economists, is that prices are not accurately capturing the cost that
oil, coal, and natural gas impose on the environment and ultimately on humanity. Producers are not paying for the cost of externalities, which
means that consumers are not charged enough and therefore consume in
too great a quantity. The solution, many economists maintain, is to force
firms to internalize these costs either by making them pay taxes on their
use of fossil fuels or by creating a system of cap and trade, both of which
require the government to step in. Some carbon taxes have already been
implemented, but they have been far too low to make a real difference. One
economist proposes the solution that the government should adjust the
rate in proportion to global temperature increases. In the case of cap and
trade, the government sets a cap on how much emission is allowed and then

issues tradeable carbon emission rights. This gives firms an incentive to use more energy-efficient technologies, so that they can sell their pollution permits to others, a strategy the carmaker Tesla has successfully employed recently. While the Environmental Defense Fund credits cap and trade for the reduction of sulfur dioxide in the atmosphere, which led to a drastic decline in acid rain, many economists acknowledge that because of intense lobbying by the oil, gas, and coal sectors, governments around the globe have allowed for too many exceptions and loopholes.[10] The International Monetary Fund suggests that, to reach net-zero emission by 2050, along with the implementation of state-financed carbon capture technologies, it is necessary to price carbon at a level that reduces emissions by 80 percent.[11] It proposes that prices be increased by 7 percent each year. This would yield relatively modest price increases in the first few years, but after a decade prices would start becoming quite prohibitive and would therefore have the intended effects of lowering demand and making other energy sources more affordable.

Many economists also remain unconvinced that the threat of global climate change is as great as environmental scientists insist. A former chief economist for the OECD, David Henderson, for example, argues that the IPCC is institutionally biased toward pessimism and lacks the proper expertise to estimate the economic costs of climate change. Also reluctant to accept the findings and suggestions of climate scientists, Nobel laureate William Nordhaus challenges the methods and assumptions employed in key environmental reports. Cambridge economist Diane Coyle further argues that the IPCC, although backed by almost all climate scientists, "is not sufficiently transparent, has not engaged effectively with critics, and lacks political legitimacy."[12] The most common complaint, however, lodged by economists against environmental scientists is that they do not properly consider the power of substitutability.

Belatedly, a few economists have begun to recognize the severity of the threat that economic growth poses to the ecosystem and the role that economics have historically played in promoting maximum exploitation of natural resources and infinite economic growth. They recognize that the projections made by environmental scientists in the past, instead of being too pessimistic, have not been dire enough. In a 2021 publication, *The Economics of Biodiversity,* Cambridge economist Partha Dasgupta addresses the economists' tradition of treating the biosphere as external to

the human economy, despite the fact that humanity has always been embedded in nature.[13] At the heart of his critique is the widespread use of GDP
to judge economic performance. GDP measures the total market dollars
of output per year, but it does not take into account the depreciation of
assets—human, capital, and nature. As a result of GDP bolstering the
focus on quantitative economic growth, it is a singularly inappropriate
device to assess the goal of sustainable economic growth. In its place,
Dasgupta suggests that economists ought to use the concept of "inclusive
wealth," which captures all of the economy's assets, including produced
capital, human capital, and natural capital.[14] The latter category, defined
to be as expansive as possible, includes everything from soils, plants, pol-
linators, and ocean currents to the global climate. Dasgupta correctly
focuses the attention not on the scarcity of specific resources but on the
capacity of the biosphere to regenerate itself. Without the hydrological,
carbon, and nitrogen cycles, life on earth would be impossible, and without sufficient biodiversity the ecological system would lose its resilience.
When an investment project is assessed within Dasgupta's proposed
framework, the criterion is not whether it adds to economic growth (that
is, GDP), but rather whether it advances inclusive wealth. If it is estimated
that a project will add to produced capital, but at the same time impose significant damage on the environment, the net effect is negative and the
project is therefore not undertaken.

Dasgupta offers plenty of specific advice on how to tackle the looming crisis. He calls for restructuring consumption and production, mas-
sively reducing waste, increasing efficiency with various technological
advances, ending subsidies that encourage overextraction and overhar-
vesting of the biosphere, implementing pollution taxes, charging resource
extraction fees, establishing protected areas, rewilding natural environments, encouraging socially responsible consumption, nudging people
toward more sustainable behavior, and developing carbon-capture technology. These efforts, he correctly argues, cannot be undertaken on the
margin, but must involve colossal endeavors on the scale of the Marshall
Plan. Together with a rethinking of the purpose of the economy, these
transformations can go a long way toward creating a sustainable economy.
The key, Dasgupta insists, is to recognize that the human economy is intrinsically bounded and that, regardless of how ingenious humanity may
be, there are limits to how much of nature can be transformed into goods

and services. The cornucopian dream of infinite substitutability that economists have held on to for so long must be recognized for what it truly is: a fantasy.

While Dasgupta's approach is not perfect, as he himself acknowledges, it constitutes a much more rigorous and responsible approach to economic development than the traditional pursuit of ever higher GDP. Dasgupta recognizes that there are massive challenges associated with accurately measuring the stock of natural capital and the damages incurred from economic activities. Such problems notwithstanding, it is far better to work with rough figures, he argues, than simply "ignore whole swathes of capital goods by pretending they do not exist."[15] Dasgupta also acknowledges that his entire approach to the economics of biodiversity is conducted in anthropocentric terms. He justifies this by arguing that nature should be "protected and promoted even when valued solely for its uses to us."[16] But once we consider that nature also has an intrinsic right to exist that extends far beyond human use, we gain even more robust reasons for protecting it. The problem is that economic reasoning has been detached from nature for too long and that it has facilitated a perception that humans are external to nature and that rich societies are independent of their poorer counterparts. But this does not imply that economics is necessarily fundamentally flawed, he argues. The problem is not with economics per se, but with how economists "have chosen to practise it."[17]

Dasgupta's revisionist approach is a crucial step toward reforming mainstream economics, but whether his views will gain traction within the profession as well as in the halls of power is an open question.

The Uses of the Past

Climate scientists warn that we now have only radical options before us: either fossil fuel growth remains dominant with dire consequences for the habitability of the earth or we reorient the economy and politics toward a new social order that will keep us within the safe operating space of the earth system. Business as usual will bring disaster. With the stakes so high, some readers might well wonder about the wisdom of our historical approach. Facing such an unprecedented and serious situation, why look to the past for explanatory frameworks or alternative values? Why not simply jettison entirely all the baggage of history and start anew?

Although it is easy to sympathize with the anger and grief that drive some people to a wholesale rejection of the past, erasing all that came before would be a catastrophic mistake, depriving us of critical knowledge while actually exacerbating the problem of Planetary Scarcity. Without the benefit of historical perspective, people are far more likely to mistake an ideological position for a universal and timeless truth. In the case at hand, the idea of Neoclassical Scarcity insists that the human condition is permanently caught between insatiable wants and limited means, with infinite substitutability providing the source of endless economic growth. Yet, far from reflecting some universal or natural truth, this notion arose from a peculiar understanding of nature, psychology, and the economy. Through some historical detective work, we have traced the emergence and descent of this idea back from present-day neoclassical economics to the marginalists of the late nineteenth century and before them the Enlightenment *philosophes,* all the way back to the defense of insatiable desire and godlike mastery of nature among seventeenth-century natural philosophers and alchemists. In this sense, our book offers a genealogical approach to historical knowledge. We show how a widely accepted normative principle came into being in a specific historical process marked by contest and conflict rather than the rational discovery of universal truth. By uncovering the hidden history of scarcity, we also begin to understand how this idea constrains our vision of the future and obscures alternative ways of seeing the economy.[18] Only by recognizing the historical specificity of Neoclassical Scarcity can we begin the search in earnest for theoretical frameworks that are better suited to guide us as we tackle the challenges brought on by the Anthropocene.

Yet our argument is not confined to a purely negative and critical approach whose sole aim is to purge destructive ideas from scholarship. Historical investigation can enrich the social sciences in far more profound ways, by letting us escape the tyranny of the present and by broadening the horizon of intellectual and political possibility. Thinking historically, we also become more adept at dealing with a complex and contingent future. In tracing the history of scarcity, we have uncovered a family tree of alternative interpretations of the relationship between nature and economy. By investigating the concepts and aims that have guided past thinkers, from David Hume to Rachel Carson, we have sought to expand and enrich the horizons of social analysis. In reconstructing the historical debates that accompanied the making of the modern world, we also un-cover paths not taken. In this sense, the past forms a storehouse of lost ideas and forgotten questions.[19] Beneath surface appearances of variation and complexity, we can detect underlying patterns that persist across time. Such archeological excavation uncovers continuities that reach all the way forward to the present moment. For example, one of the main findings of our book is the deep and growing influence of cornucopian thought from the seventeenth century to the present. Yet, our analysis also reveals coun-termovements and positions of resistance. For example, the legacy of romantic thought has remained a potent influence on the opponents of cornucopianism from the late eighteenth century to the present. Like-wise, socialist critiques of capitalism, with roots dating back at least to Thomas More's *Utopia* in the sixteenth century, have shown remark-able vitality and perseverance.

We should not be surprised then to see how the storehouse of the past shapes the current moment. In searching for alternatives to how neoclas-sical economics theorizes the nature-economy nexus, thinkers from across the political spectrum have turned to a wide range of past ideas and ide-ologies for inspiration. Some critics have eschewed European intellectual traditions altogether in favor of non-Western systems of thought, but many still look to the concepts we have excavated in this book. Pope Francis's encyclical *Laudato si,* a devastating critique of modern consumerism and environmental degradation, revives the Christian ideal of curbing desire and living within limits. The American environmentalist Bill McKibben, founder of the 350 Movement, embraces notions of self-sufficiency and degrowth that harken back to Rousseau and the Romantics. On the secu-lar left, a new generation of scholar-activists looks to Marx and the other socialists and their critique of capitalism for a deeper understanding of the origins of climate change. Some Neo-Marxists put their hope in transformative technology like geoengineering and carbon removal while others explore forms of flourishing that reconcile human welfare with ecological limits and Planetary Boundaries. By drawing on both re-cent and distant traditions, these competing movements have forged a range of creative responses to Planetary Scarcity. It is worth noting that none of these efforts are simply reactionary or nostalgic; all seek to adapt and transform the worlds of the past to the problems of the present. Such rival responses in turn mirror the fractured condition of humanity. Per-

sistent geopolitical divisions, ideological polarization, and cultural differences appear to preclude the possibility of a single dominant understanding of the Anthropocene.[20]

We have simplified the findings of this book by grouping the various
approaches to scarcity under two umbrella terms. The first kind involves
a family of ideas that endorses an active mastery of nature together with a
dynamic and expansive notion of desire: Cornucopian Scarcity, Enclosure Scarcity, Enlightened Scarcity, Capitalist Scarcity, and Neoclassical
Scarcity. This tradition of *Cornucopian* ideology first emerged in the
seventeenth century and eventually reached a dominant position by
the end of the nineteenth century. A second cluster of ideas in our history
revolves around limits to human power over nature and the need for constraint and moderation of human desires. This was a *Finitarian* ideology
of bounded economies rather than open frontiers. It was the dominant
worldview of sixteenth-century Neo-Aristotelian Scarcity. Later expres-
sions of Finitarianism have included Utopian Scarcity, Romantic Scarcity,
Malthusian Scarcity, and Socialist Scarcity.

The conflict between Cornucopianism and Finitarianism is still
playing out in the current moment, yet there is also growing recognition
that we simply cannot afford to rehash the same old rivalry. This duel cannot go on forever. While Cornucopians have had the upper hand until recently, the accelerating pace of growth and scale of extraction in the global
economy has ended up creating environmental problems of unprecedented
gravity at the level of the earth system, such as climate change, oceanic
acidification, and nitrogen overloading. At this point, the long struggle
between Finitarians and Cornucopians seems to have reached a new
stage. Cornucopian ideology may well persist for a long time to come,
but the planet itself now seems to weigh in on the side of the Finitarians.

Our excavation of Finitarian scarcities presents a map of possible
paths to guide new ways of thinking about the economy and nature. Early
modern Christian and Utopian thinkers conceived of the economy as circular. They imagined the economic activity of the nation and kingdom as
an orderly and bounded sphere—along the lines of an idealized family
household—embedded in the divine order of the natural world. Desire was
harnessed toward moral and spiritual ends, not insatiable consumption.
This hierarchical conception of the economy was not just expressed in religious tenets and moral maxims but also entrenched in legal restrictions

and popular tradition, including sumptuary laws to regulate consumer desire, poor laws to provide parish welfare, and customary use rights to access common land. In the early modern traditional social order, long-term
growth was neither a political objective nor a moral imperative. The purpose of human desire was not to stimulate endless new forms of consumption. While this Christian cosmology fell apart in the seventeenth and
eighteenth centuries, it left a legacy that was never wholly eradicated. Liberal economic thinkers reworked the question of limits to desire by suggesting that human needs would become saturated over time. David Hume
argued that needs and wants would undergo gradual refinement, in effect
decoupling pleasure and enjoyment from their material basis. Over time,
as minds were polished and refined, people would opt for higher pleasures,
such as conversation, poetry, and art. Indeed, even Alfred Marshall held
out the possibility for such a development. Expanding on the same theme,
John Stuart Mill and John Maynard Keynes hoped that humans would
eventually liberate themselves from material needs and occupy themselves instead with the pursuit of what they called the "Art of Life."

Among the critics of commercial society, the concept of a bounded
economy provided an essential alternative to liberal notions of expansion
and growth. Thomas More and Gerrard Winstanley wanted to restore a
need-based social order. Jean-Jacques Rousseau and John Ruskin drew
on the model of the household to imagine convivial stationary states. In
the twentieth century, the image of the circular economy gained new life,
inspired by the science of thermodynamics and systems ecology as well
as feminist theories of the household. The ecological economist Kenneth
Boulding imagined a blend of affluence and austerity in the closed spaceship economy. Ecofeminists saw in the maintenance of the household the
true locus of human welfare, recentering the economy toward care work
and reproduction. Coming full circle back to the early moderns, the feminist theorist and environmental historian Carolyn Merchant recovered
ancient cosmologies of the nurturing principle of nature (a theological
variation on the care principle of the household) to attack the mechanistic and patriarchal origins of cornucopianism.

Such a reorientation requires a new scalar imagination: we need to
think on the scale of the earth system *while* taking a long-term approach to
the economy in the name of intergenerational equity. As we have seen, Cornucopian and Finitarian approaches differ markedly in how they view the

future. The idea of insatiable desire presumes a specific conception of time and temporality. Cornucopian philosophers and agricultural improvers of the seventeenth century imagined a rapid transition toward earthly abundance—what they called the Great Instauration—achieved through the deciphering of nature's source code, designed by God, the Creator. Later versions of cornucopianism retained this profound optimism, but grew more circumspect about the precise content of the future. When traditional Christian cosmology began to lose ground in the Enlightenment, the future became the territory of competing political and social interpretations. Paradoxically, such faith in future progress often precluded actual long-term thinking. While the advocates of neoclassical growth theory expect a dynamic future, predicated on continuous technical innovation, they show little interest in questions of intergenerational solidarity or long-term social developments. The future takes the form of investment decisions big and small. Prudential firms and individuals weigh present cost and future benefit with an eye to the discount rate. We find a parallel reticence about the future in Marx, whose vision of socialism promised a total rupture with the capitalist system but offered little in the way of a blueprint for the political and social order after the Revolution.

In contrast, Finitarian forms of scarcity have produced more specific recipes for long-term thinking. We can understand the circular economy of early modern Christian and Utopian thought as a strategy to preserve enduring stability and encourage spiritual rather than material wants—what Roman Krznaic terms "cathedral thinking."[21] Such forms of thought did not become extinct in modern times. For example, John Ruskin saw the exhaustion of coal as a source of hope and renewal, necessitating a return to skilled labor and a circular economy. Ruskin drew on medieval architecture and art to imagine the possibility of intergenerational flourishing after fossil fuel. John Stuart Mill also looked forward to a future steady state during which the social strife intrinsic to the growth phase was eliminated and people could peacefully pursue the quest for the good life. Fourier had an even more ambitious vision of the future. He described in great detail how people might go about fundamentally redirecting their desires, away from excessive consumption toward libidinal and libertine pleasures. Such concerns about future possibilities continued to surface in the twentieth century. For Keynes, advancements in the science and technology raised the possibility of a fundamental shift in the economy:

How should people plan for the end of work and the coming of abundance? In Arendt's critique of affluent society, the acceleration and multiplication of consumer wants was a corrosive force, wearing down the space for political action and reducing citizens into drones. She, too, looked to art and architecture to defend the significance of the long term. Without the persistence of a durable world of art and action, humans became trapped in a myopic present, captive to their own desires.

The central theme of this book has been to explore the manifold ways in which humans have imagined their relation with nature. Neoclassical Scarcity rests on the idea of humans as intelligent artificers—*Homo faber*—capable of remaking the world in their image. This dualistic and mechanistic worldview attributes intelligence only to humans who give form to matter by mastering the passive and inert resources of the world on the basis of mechanistic principles. While this kind of dualism has been crucial in sustaining Cornucopian ideology since the seventeenth century, it has never been entirely dominant. A major alternative idea of nature relies not on mechanistic science but the respectful mimicry of organic processes. From this perspective, humans learn from nature how to make useful tools, without assuming absolute mastery. In fact, this approach tends to endow nature with an intrinsic force and complexity that humans may emulate to some degree, but where nature takes the lead and people simply follow. We find a variation on this theme already in eighteenth-century vitalist philosophy and physiocracy, which saw power over nature not in terms of absolute mastery but a collaborative partnership. Here, human labor contributed some part of the overall value of the product together with the work of nature.

This notion of partnership was common among the classical political economists, including the Physiocrats and Smith. There was an echo of it also in Marx, though his theory was almost entirely devoted to the human component of the partnership rather than its ecological foundation. However, with the marginalist economists, the partnership of man and nature lost ground to a fundamentally anthropocentric notion of value. As agricultural production grew in efficiency, nature forfeited its central place in the economy and the theories of economists. Yet in the same historical moment, the idea of partnership received a dramatic new expression in natural science. For Darwin, the human power of breeding (what he called "artificial" or "methodical" selection) amounted to a pale imitation of the superior ingenuity and rationality of natural selection. In the twentieth century, this pessimistic view of human capacity became the foundation for a new ecological critique of capitalism. Rachel Carson argued that humans were meddling like clumsy children in systems they could manipulate but not fully understand. Where Darwin had seen little reason to worry about human intervention, Carson thought the unintended consequences of interference might destroy the web of life. Humans behaved like lords of all creation, yet in practice they had become a destructive parasite on the life process. As we have seen, this threat has only grown in scope since Carson's death. The Great Acceleration now disturbs the basic biogeochemical processes that keep the earth hospitable to complex societies.

New approaches to Planetary Scarcity also require us to confront the problem of distribution and equality from the perspective of the earth system. Modern economics assumes that perpetual growth can legitimate the social order even if relative inequality persists. Poor countries and lower classes might lag behind the rich, but as long as technological innovation proceeds apace, living standards across the globe will continue to improve. What will happen to the global order if this promise of rising standards turns out to be false? The new science of Planetary Scarcity underscores the finite capacity of the earth to absorb the waste products of the global economy. It also insists that a stable climate and biodiversity provide the biophysical foundation for all economic activity. We are at a moment when globalization is putting increasing pressure on sinks, resources, and biodiversity. Since business as usual will drive the global economy toward multiple tipping points, one might reasonably conclude that the promise of perpetual growth cannot be made a universal standard for the whole human population. Imagine the effect on world politics if this discovery became common knowledge. At the moment, earth system science does not look like a vehicle for radical social change. And yet, we may one day look back at climate science and the ecology of biodiversity as the catalysts that led to the transcendence of the economic order of the twentieth century.[22]

The end of the idea of perpetual growth would challenge not just Neoclassical Scarcity but also Socialist Scarcity. In Marx's framework, the prospect of socialism was explicitly tied to the pursuit of large-scale industry and agriculture, fueled by steam and coal. While some socialists have aspired for an ecological understanding of justice, much of the movement still celebrates growth and mastery as preconditions of equality. Yet as Rousseau reminds us, egalitarianism can be defended on different grounds, beginning with material simplicity and republican liberty rather than Promethean industrial technology. This alternative conception of the good life might make it possible to imagine a process of global convergence around new standards of human flourishing. The effort to raise material standards in the Global South would have to be accompanied by a concomitant reduction of the ecological footprint for the affluent countries. Such convergence would require a new model of development, which favored material growth only up to a certain universal income threshold. Growth in one region or class would have to be balanced by degrowth or a lower rate of material growth in another.[23]

Slowing down the Great Acceleration will require immense effort and creativity in cultural and technological terms. The complex challenge of Planetary Scarcity rules out a strategy that focuses *only* on restraint and withdrawal. Humanity faces threats so dangerous that extensive technological intervention and cultural change have become necessary. To back off, we first need to repair what has been damaged. The Great Deceleration is not a moment for technophobia or apocalyptic pessimism. Our best bet may be to exorcise Cornucopianism from culture and ideology and replace it with a new politics and technology of repair, oriented toward the goal of universal flourishing within planetary constraints.

This radical future of repair extends to many different domains. Fossil fuel economies must transition to renewable energy while at the same time removing carbon from the atmosphere. In technical terms, this might involve projects of carbon sequestration of different kinds, through the constructing of artificial sinks by means of underground storage or the enhancement of the ocean's capacity to absorb carbon. Such carbon removal industries could provide employment and livelihoods in places where meaningful work is hard to come by. Yet, at the moment carbon sequestration technologies are very expensive and energy intensive. Repair might also take the political form of debt payment where the affluent countries pledge to clean up the mess they have made while assisting goals of just development in the Global South. By halting land use change and repairing ecosystems, humans might regenerate and even expand the natural carbon sinks of the earth. Some critics imagine a planetwide project of rewilding to this effect. This would make room for functional and genetic biodiversity by setting aside sufficient space where nonhuman life forms could thrive. Here again, political priorities will determine the shape of things to come—should a vegan diet become a global priority to save land for biodiversity?[24]

Curbs on land use would also prevent the spread of new pathogens. Forest logging, road construction, and other points of contact between humans and nonhuman species provide key pathways for the emergence of new infectious diseases. Virologists predict that the frequency of epidemics will only increase as commodity frontiers expand. Here, an ethos of ecological repair would not only preserve the integrity of the natural world but also help restore the health and welfare of human beings, especially in those social classes and minority populations most vulnerable to epidemics. In part, this is a question of balancing local livelihoods with conservation aims in the Global South; in part, it is a problem of limiting or reorienting consumption in affluent countries.

Without a doubt, the ethos of repair will require a profound psychological shift across the planet. Cornucopian ideology has usurped universal aspirations of equality, freedom, and creative fulfillment. Any viable alternative to the present order needs to come to terms with the human desire for self-determination and find means to channel it in new directions. Repair does not rule out an element of dynamism. Circular economies can still foster individual and collective forms of creativity, pleasure, and play, channeled through science and art as well as everyday living. But whatever direction freedom takes in the Great Deceleration, human desire will be bounded by the new condition of Planetary Scarcity.

NOTES

Introduction

1. Simon Lewis and Mark Maslin Simon, *The Human Planet: How We Created the Anthropocene* (New Haven: Yale University Press, 2018); Julia Adeney Thomas, Mark Williams, and Jan Zalasiewicz, *The Anthropocene: A Multidisciplinary Approach* (London: Polity, 2020).

2. The idea of nature has a long and complex history. For some hints about this lineage, see Raymond Williams, "Ideas of Nature," in *Problems in Materialism and Culture: Selected Essays* (London: Verso, 1980), 67–85.

3. Will Steffen et al., "Planetary Boundaries: Guiding Human Development on a Changing Planet," *Science* 347, no. 6223 (February 13, 2015); for COVID-19, see Rory Gibb et al., "Zoonotic Host Diversity Increases in Human-dominated Ecosystems," *Nature* 584, no. 7821 (August 20, 2020): 398–402.

4. Marshall Sahlins, *Stone Age Economics* (1972; London: Routledge, 2017), 11.

5. Shaun A. Marcott, Jeremy D. Shakun, Peter U. Clark, Alan C. Mix, "A Reconstruction of Regional and Global Temperature for the Past 11,300 Years," *Science* 339 (March 8, 2013): 1198–1201; Franz Mauelshagen, "The Dirty Metaphysics of Fossil Freedom," in *The Anthropocenic Turn: The Interplay between Disciplinary and Interdisciplinary Responses to a New Age,* ed. Gabriele Dürbeck and Philip Hüpkes, 59–76 (New York: Routledge, 2020).

6. *Oxford English Dictionary,* 2nd ed. (Oxford: Oxford University Press, 2004), s.v. "Scarcity."

7. Strother Roberts, *Colonial Ecology, Atlantic Economy: Transforming Nature in Early New England* (Philadelphia: University of Pennsylvania Press, 2019), 120–125.

8. David Hume, "Of the Populousness of Ancient Nations," in *Essays, Moral, Political, and Literary,* ed. Eugene Miller (1777; Indianapolis: Liberty Fund, 1985), 451.

9. Will Steffen, Paul Crutzen, and John McNeill, "The Anthropocene: Are Humans Now Overwhelming the Great Forces of Nature?" *Ambio* 36, no. 8 (2007), 617, table 1.

10. Heinz Kohler, *Intermediate Microeconomics: Theory and Applications* (Glenview, IL: Scott Foresman, 1982).

11. Lionel Robbins, *An Essay on the Nature and Significance of Economic Science* (1932; New York: New York University Press, 1984), 15.

12. Paul Crutzen and Eugene Stoermer, "The Anthropocene," *IGBP Newsletter* 41 (2000): 17–18; Johan Rockström et al., "A Safe Operating Space for Humanity," *Nature* 461 (2009): 472–475; Steffen et al., "Planetary Boundaries."

13. Will Steffen et al., "Trajectories of the Earth System in the Anthropocene," *Proceedings of the National Academy of Sciences* 115, no. 33 (2018): 8252–8259. For the definition of sinks, see Brian J. Skinner and Barbara W. Murck, *The Blue Planet: An Introduction to Earth System Science,* 3rd ed. (Hoboken, NJ: John Wiley and Sons, 2011).

14. Will Steffen et al., "The Anthropocene: From Global Change to Planetary Stewardship," *Ambio* 40, no. 7 (2011): 739–761.

15. Tibor Scitovsky, *The Joyless Economy: An Inquiry into Human Satisfaction and Consumer Dissatisfaction* (Oxford: Oxford University Press, 1976); Richard Easterlin, "Does Economic Growth Improve the Human Lot? Some Empirical Evidence," in *Nations and Households in Economic Growth,* ed. Paul David and Melvin Reder, 89–125 (New York: Academic Press, 1974).

16. Fred Hirsch, *Social Limits to Growth* (Cambridge, MA: Harvard University Press, 1976).

17. Amartya Sen, *Development as Freedom* (New York: Anchor Books, 1999), 5.

18. Karen Raworth, *Doughnut Economics: Seven Ways to Think Like a 21st Century Economist* (White River Junction, VT: Chelsea Green, 2017), 30.

19. Partha Dasgupta, *The Economics of Biodiversity: The Dasgupta Review* (London: HM Treasury, 2021).

20. A keyword search across more than 15,000 PhD dissertations written at 152 US and Canadian universities between 2001 and 2018 reveals very little interest in climate change among young scholars. Only 0.5 percent of disser-

tations included climate change or global warming as keywords. Michael Roos and Franziska M. Hoffart, "Importance of Climate Change in Economics," in *Climate Economics: A Call for More Pluralism and Responsibility,* 19–34 (Cham, Switzerland: Palgrave, 2021), 23, 29.

21. John Maynard Keynes, *The General Theory of Employment, Interest, and Money* (London: Macmillan, 1936), 383.

22. Terry Eagleton, *Ideology: An Introduction* (London: Verso, 1996), 52.

23. In so doing, we follow in the tradition of previous scholars who have explored the history of the concept of scarcity, including Hugo Hegeland, *Från Knapphet till Överflöd: En studie över knapphetsbegreppet i nationaleckonomi* (Stockholm: Nature och Kultur, 1967); Nicholas Xenos, *Scarcity and Modernity* (London: Routledge, 1989); Regina Gagnier, *The Insatiability of Human Wants* (Chicago: University of Chicago Press, 2000); Adel Daoud, *Scarcity, Abundance, and Sufficiency: Contributions to Social and Economic Theory* (Gothenburg: University of Gothenburg Press, 2011); Costas Panayotakis, *Remaking Scarcity: From Capitalist Inefficiency to Economic Democracy* (London: Pluto, 2011); Gustav Peebles, "For a Love of False Consciousness: Adam Smith on the Social Origins of Scarcity," *Economic Sociology* 12, no. 3 (2011): 19–25; and Andrew Abbott, "The Problem of Excess," *Sociological Theory* 32, no. 1 (2014): 1–26. Our book also builds on scholarship exploring the long-term dynamics between nature and economy, including Margaret Schabas, *The Natural Origins of Economics* (Chicago: University of Chicago Press, 2005); Donald Worster, *Shrinking the Earth: The Rise and Decline of Natural Abundance* (Oxford: Oxford University Press, 2016); Nathaniel Wolloch, *Nature in the History of Economic Thought: How Natural Resources Became an Economic Concept* (London: Routledge, 2017); Paul Warde, Libby Robin, and Sverker Sörlin, *The Environment: A History of the Idea* (Baltimore: Johns Hopkins University Press, 2018); and Pierre Charbonnier, *Affluence and Freedom: An Environmental History of Political Ideas* (London: Polity, 2021).

24. See, for example, Vandana Shiva, *Staying Alive: Women, Ecology and Survival in India* (London: Zed Books, 1988); Ramachandra Guha, *Environmentalism: A Global History* (New York: Longman, 2000); Md Saidul Islam, "Old Philosophy, New Movement: The Rise of the Islamic Ecological Paradigm in the Discourse of Environmentalism," *Nature and Culture* 7, no. 1 (2012): 72–94; Robin Wall Kimmerer, *Braiding Sweetgrass: Indigenous Wisdom, Scientific Knowledge, and the Teaching of Plants* (Minneapolis: Milkweed, 2013); and Kimberly K. Smith, *African American Environmental Thought* (Lawrence: University Press of Kansas, 2021).

25. Dipesh Chakrabarty underscores the centrality of reverence mixed with fear to any genuine understanding of our planetary condition. See Dipesh Chakrabarty, *The Climate of History in a Planetary Age* (Chicago: University of Chicago Press, 2021), 198.

1. Types of Scarcity before 1600

1. John L. Brooke, *Climate Change and the Course of Global History: A Rough Journey* (Cambridge: Cambridge University Press, 2014), 383.

2. Thomas Hobbes, *Leviathan* (1651; Cambridge: Cambridge University Press, 1991), 89.

3. John Walter and Keith Wrightson, "Dearth and the Social Order in Early Modern England," *Past and Present* 71, no. 1 (1976): 22–42, 22.

4. Quoted in Carlo Cipolla, *Before the Industrial Revolution: European Society and Economy, 1000–1700* (New York: Norton, 1976), 159.

5. Edmund Dudley, *The Tree of Commonwealth*, ed. D. M. Brodie (1510; Cambridge: Cambridge University Press, 1948), 90–91.

6. Neal Wood, *Foundations of Political Economy: Some Early Tudor Views on State and Society* (Berkeley: University of California Press, 1994), 74.

7. Dudley, *The Tree of Commonwealth*, 44.

8. Laurentius Petri, *Oeconomia Christiana*, selections reprinted in Anders Björnsson and Lars Magnusson, eds., *Jordpäron: Svensk Ekonomihistorisk Läsebok* (Stockholm: Atlantis, 2011), 95. Authors' translation of quoted passage.

9. Joel Kaye, *Economy and Nature in the Fourteenth Century: Money, Market Exchange, and the Emergence of Scientific Thought* (Cambridge: Cambridge University Press, 1998), 61.

10. Wood, *Foundations of Political Economy*, 80.

11. Per Brahe, *Oeconomia eller Hushållsbok för ungt adelsfolk* (Economy, or household book for young nobles), reprinted in Björnsson and Magnusson, *Jordpäron*, 95–96. Authors' translation of quoted passage.

12. Eamon Duffy, *The Voices of Morebath: Reformation and Rebellion in an English Village* (New Haven: Yale University Press, 2003).

13. Duffy, *Voices of Morebath*, 11.

14. A. L. Morton quoted in Christopher Kendrick, *Utopia, Carnival and Commonwealth in Renaissance England* (Toronto: University of Toronto Press, 2004), 77.

15. Craig Muldrew, *Food, Energy and the Creation of Industriousness: Work and Material Culture in Agrarian England, 1550–1780* (Cambridge: Cambridge University Press, 2011), 46.

16. Duffy, *Voices of Morebath*, 16–32.

17. Carlo Ginzburg, *The Night Battles: Witchcraft and Agrarian Cults in the Sixteenth and Seventeenth Centuries* (Baltimore: Johns Hopkins University Press, 2013); Carlo Ginzburg, *Ecstasies: Deciphering the Witches' Sabbath* (Chicago: University of Chicago Press, 1991).

18. Jayne Elisabeth Archer, Richard Marggraf Turley, and Howard Thomas, "The Autumn King: Remembering the Land in 'King Lear,'" *Shakespeare Quarterly* 63, no. 4 (Winter 2012): 518–543.

19. Francis Bacon, "A Brief Discourse Touching the Happy Union of the Kingdoms of England and Scotland: Dedicated in Private to His Majesty," in *The Letters and the Life of Francis Bacon,* ed. James Spedding, 7 vols. (London: Longmans, Green, Reader, and Dyer, 1868), 3: 90.

20. Aristotle, *Nichomachean Ethics,* trans. Martin Ostwald (New York: Macmillan, 1962).

21. Aristotle, *The Politics,* Stephen Everson, ed. (Cambridge: Cambridge University Press, 1988), 13.

22. Aristotle, *Politics,* 14.

23. Brad Gregory, *The Unintended Reformation: How a Religious Revolution Secularized Society* (Cambridge, MA: Harvard University Press, 2012), 262–263.

24. Luther, *On Commerce and Usury (1524),* ed. Philipp Robinson Rössner (London: Anthem, 2015), 175.

25. Luther, *On Commerce and Usury,* 177.

26. Luther, *On Commerce and Usury,* 178.

27. Erasmus shared the same sentiments. He chastised the merchants for hunting "for profits from every possible source, by trickery, by lies, by fraud, by cheating, buying up here what they sell for more than double there, and robbing the wretched poor with their monopolies." Quoted in Margot Todd, *Christian Humanism and the Puritan Social Order* (Cambridge: Cambridge University Press, 1987), 128.

28. Luther, *On Commerce and Usury,* 184.

29. Luther, *On Commerce and Usury,* 184–191.

30. Luther, *On Commerce and Usury,* 190.

31. Luther, *On Commerce and Usury,* 173.

32. Jacques Le Goff, in *Your Money or Your Life: Economy and Religion in the Middle Ages* (Cambridge: Zone Books, 1990), offers an insightful discussion of how the church altered its message to enable merchants to conduct their affairs.

33. Ethan Shagan, *The Rule of Moderation: Violence, Religion and the Politics of Restraint in Early Modern England* (Cambridge: Cambridge University Press, 2011), 35.

34. Shagan, *Rule of Moderation,* 32.

35. Matthew Ingleby and Samuel Randalls, *Just Enough: The History, Culture and Politics of Sufficiency* (London: Palgrave, 2019).

36. Martha Howell, *Commerce before Capitalism in Europe, 1300–1600* (Cambridge: Cambridge University Press, 2010), 208.

37. Acts of the Privy Council, Enforcing Statutes of Apparel, issued at Greenwich, 15 June 1574, 16 Elizabeth I.

38. Ulinka Rublack and Giorgio Riello "Introduction," in *The Right to Dress: Sumptuary Laws in a Global Perspective, c. 1200–1800,* ed. Riello and Rublack, 1–34 (Cambridge: Cambridge University Press, 2019), 16–19.

39. Keith Wrightson, *Earthly Necessities: Economic Lives in Early Modern Britain* (New Haven: Yale University Press, 2000); and Jo Guldi, *The Long Land War: The Global Struggle for Occupancy Rights* (New Haven: Yale University Press, 2022).

40. Thomas More, *Utopia,* ed. George Logan and Robert Adams (1516; Cambridge: Cambridge University Press, 1989), 18–19.

41. More, *Utopia,* 18–19.

42. More, *Utopia,* 19.

43. More, *Utopia,* 39.

44. More, *Utopia,* 40.

45. More, *Utopia,* 51.

46. More, *Utopia,* 77.

47. Seneca, *Letters from a Stoic,* trans. Robin Campbell (London: Penguin, 1969), 34, 65, 168.

48. More, *Utopia,* 76.

49. More, *Utopia,* 109.

50. More, *Utopia,* 65.

51. More, *Utopia,* 38.

52. More, *Utopia,* 52; Aristotle, *Ethics,* 125.

53. More, *Utopia,* 109.

54. Seneca, *Dialogues and Letters,* ed. and trans. C. D. N. Costa (London: Penguin, 1997), 42, 104.

2. Cornucopian Scarcity

1. Geoffrey Parker, "Crisis and Catastrophe: The Global Crisis of the Seventeenth Century Reconsidered," *American Historical Review* 113, no. 4 (2008): 1053–1079.

2. Giovanni Botero, *On the Causes of the Greatness and Magnificence of Cities,* trans. Geoffrey Symcox (Toronto: University of Toronto Press, 2000), 42. Botero used the term *accrescere,* derived from *crescita,* which means growth.

3. Botero, *Causes of the Greatness,* 44.

4. Deborah E. Harkness, *The Jewel House: Elizabethan London and the Scientific Revolution* (New Haven: Yale University Press, 2007), 2.

5. Francis Bacon, *The New Organon,* ed. Lisa Jardine and Michael Silverthorne (Cambridge, UK: Cambridge University Press, 2000), 7.

6. Bacon, *New Organon,* 13.

7. Bacon, *New Organon,* 223.

8. Bacon, *New Organon,* 102.

9. Pierre Hadot, *The Veil of Isis: An Essay on the History of the Idea of Nature,* trans. Michael Chase (Cambridge, MA: Harvard University Press, 2008); Reinhart Koselleck, *Futures Past: On the Semantics of Historical Time* (New York: Columbia University Press, 2004).

10. Bacon, *New Organon,* 99.

11. Bacon, *New Organon,* 65–66.

12. Francis Bacon, *New Atlantis,* in Susan Bruce, ed., *Three Early Modern Utopias: Utopia, New Atlantis, and The Isle of Pines* (Oxford: Oxford University Press, 1999), 177–178.

13. Max Horkheimer and Theodor Adorno, *The Dialectic of Enlightenment* (New York: Continuum, 1994).

14. Bacon, *New Atlantis,* 183.

15. Bacon, *New Atlantis,* 185.

16. Paul Slack, *The Invention of Improvement: Information and Material Progress in Seventeenth-Century England* (Oxford: Oxford University Press, 2015), 92.

17. John Milton, *Of Education* (London, 1644), 363.

18. Hartlib, *Samuel Hartlib and His Legacy of Husbandry* (London, 1655), 247–250; and Cressy Dymock, *A Discovery for New Divisions, or Setting out of Lands, as to the best Forme* (London, 1653), 12.

19. William Newman, *Newton the Alchemist: Science, Enigma, and the Quest for Nature's "Secret Fire"* (Princeton: Princeton University Press, 2018).

20. Samuel Hartlib, *An Essay for Advancement of Husbandry-Learning* (London, 1651), ii.

21. Samuel Hartlib, *Propositions for Advancement of Husbandry-Learning* (London, 1651), 3.

22. Carolyn Merchant, *The Death of Nature: Women, Ecology, and the Scientific Revolution* (New York: HarperCollins, 1980).

23. Francis Bacon, "Thoughts and Conclusions on the Interpretation of Nature or A Science of Productive Works," in Benjamin Farrington, *The Philosophy of Francis Bacon: An Essay on Its Development from 1603 to 1609* (Liverpool: Liverpool University Press, 1964), 99.

24. Merchant, *Death of Nature,* 169–172.

25. Gabriel Plattes, *A Description of the Famous Kingdome of Macaria* (London, 1641), 2.

26. Plattes, *Famous Kingdome of Macaria,* 4.

27. Plattes, *Famous Kingdome of Macaria,* 5.

28. Plattes, *Famous Kingdome of Macaria,* 11.

29. Gabriel Plattes, *A Discovery of Infinite Treasure, Hidden since the World's Beginning* (London, 1639), iv.

30. Gabriel Plattes, *The Profitable Intelligencer, Communicating His Knowledge for the Generall Good of the Common-Wealth and All Posterity* (London, 1644), 3.

31. Thomas Hobbes, *Leviathan,* ed. Richard Tuck (1651; Cambridge: Cambridge University Press, 1991), 89.

32. Ellen Meiksins Wood, *The Origin of Capitalism: A Longer View* (New York: Verso, 1999), 106.

33. Henry Robinson, *Certain Proposalls in order to the Peoples Freedome and Accomodation in some Particulars with the Advancement of Trade and Navigation of this Commonwealth in Generall* (London: M. Simmons, 1652), 18. For further discussion, see Carl Wennerlind, *Casualties of Credit: The English Financial Revolution, 1620–1720* (Cambridge, MA: Harvard University Press, 2011).

34. Christopher Hill, *The World Turned Upside Down: Radical Ideas during the English Revolution* (London: Penguin, 1984).

35. Gerrard Winstanley, *A Declaration from the Poor Oppressed People of England* (1649), 1.

36. Winstanley, *Poor Oppressed People of England,* 1.

37. Winstanley, *Poor Oppressed People of England,* 1.

38. Winstanley, *Poor Oppressed People of England,* 2.

39. Winstanley, *Poor Oppressed People of England,* 2.

40. Winstanley, *Poor Oppressed People of England,* 4.

41. John Locke, *Two Treatises of Government* (Cambridge: Cambridge University Press, 1960), 290.

42. Locke, *Two Treatises of Government,* 291.

43. Locke, *Two Treatises of Government,* 294.

44. Locke, *Two Treatises of Government,* 294.

45. Locke, *Two Treatises of Government,* 297.

46. John Locke, "An Essay on the Poor Laws [1697]," in *Political Essays,* ed. Mark Goldie (Cambridge: Cambridge University Press, 1997), 185–187.

47. Locke had a documented interest in Paracelsian alchemy and clearly embraced Baconianism, yet he did not place as much importance on scientific advancement as members of the Hartlib Circle. For him, labor was the primary foundation for economic improvement. For a discussion of Locke on alchemy, see Guy Meynell, "Locke and Alchemy: His Notes on Basilius Valentinus and Andreas Cellarius," *Locke Studies* 2 (2002): 177–197.

48. John Locke, *Essay on Human Understanding,* ed. Peter Nidditch (1689; Oxford: Clarendon Press, 1979), 262–263.

49. Andrea Finkelstein, "Nicholas Barbon and the Quality of Infinity," *History of Political Economy,* 32, no. 1 (2000): 83–102, 97.

50. Barbon, *A Discourse of Trade* (London, 1690), 5.

51. Barbon, *A Discourse of Trade,* 18–19. See Steve Pincus and Alice Wolfram, "A Proactive State? The Land Bank, Investment and Party Politics in the 1690s," in *Regulating the British Economy, 1660–1850,* ed. Perry Gauci (Farnham: Ashgate, 2011), 42.

52. Pierre Force, *Self-Interest before Adam Smith: A Genealogy of Economic Science* (Cambridge: Cambridge University Press, 2007).

53. Slack, *The Invention of Improvement,* 154.

54. Jean-Louis Flandrin, "Introduction: The Early Modern Period," in *Food: A Culinary History,* ed. Jean-Louis Flandrin, Massimo Montanari, and Albert

Sonnenfeld, trans. Clarissa Botsford (New York: Columbia University Press, 1999), 361.

55. Jan de Vries, *The Industrious Revolution: Consumer Behavior and the Household Economy, 1650 to the Present* (Cambridge: Cambridge University Press, 2008).

56. Joyce Appleby, "Ideology and Theory: The Tension between Political and Economic Liberalism in Seventeenth-Century England," *American Historical Review* 81, no. 3 (1976): 499–515, 505.

57. Barbon, *A Discourse of Trade,* 14.

58. Barbon, *A Discourse of Trade,* 15.

59. Barbon, *A Discourse of Trade,* 16.

60. The classic statement on this subject is A. O. Hirschman, *The Passions and the Interests: Political Arguments for Capitalism before Its Triumph* (Princeton: Princeton University Press, 1977). For a more recent discussion, see Jonathan Sheehan and Dror Wahrman, *Invisible Hands: Self-Organization and the Eighteenth Century* (Chicago: University of Chicago Press, 2015).

61. Barbon, *A Discourse of Trade,* 65.

62. Roger North, *The Lives of the Norths* (London, 1890), 3: 56.

63. Barbon, *A Discourse of Trade,* 6.

64. Barbon, *A Discourse of Trade,* 5.

65. Barbon, *A Discourse Shewing the Great Advantages that New-Buildings, and the Enlarging of Towns and Cities Do Bring to a Nation* (London, 1678), 1.

66. Barbon, *Great Advantages,* 2.

67. Barbon, *A Discourse of Trade,* 62.

68. Barbon, *A Discourse of Trade,* 5.

69. Barbon, *A Discourse of Trade,* 35.

70. For other voices similarly defending conspicuous consumption, see Paul Slack, "The Politics of Consumption and England's Happiness in the Later Seventeenth Century," *English Historical Review* 122, no. 497 (2007): 609–631.

71. Roy Porter, "Consumption: Disease of the Consumer Society," in *Consumption and the World of Goods,* ed. John Brewer and Roy Porter (London: Routledge, 1993), 66.

72. Bernard Mandeville, *The Female Tatler,* November 30, 1709.

73. Bernard Mandeville, *The Fable of the Bees or Private Vices, Publick Benefits,* ed. F. B. Kaye, 6th ed. (Liberty Fund: Indianapolis, 1988), 13, 24.

74. Mandeville, *The Fable of the Bees,* 25.

75. Mandeville, *The Fable of the Bees,* 36.

76. Mandeville, *The Fable of the Bees,* 105.

77. Mandeville, *The Fable of the Bees,* 119.

78. Mandeville, *The Fable of the Bees,* 128 (emphasis added).

79. For a rich account of the competing ideas of those who believed in the prospects of infinite growth and those who insisted that finite landed wealth

must constitute the foundation of society, see Steve Pincus, *1688: The First Modern Revolution* (New Haven: Yale University Press, 2009).

80. Istvan Hont, "The Early Enlightenment Debate on Commerce and Luxury," in *The Cambridge History of Eighteenth-Century Political Thought,* ed. Mark Goldie and Robert Wokler (Cambridge: Cambridge University Press, 2008).

81. François de Salignac de La Mothe-Fénelon, *Telemachus, Son of Ulysses,* ed. and trans. Patrick Riley (Cambridge: Cambridge University Press, 1994), 297.

82. Fénelon, *Telemachus,* 296–298.

83. Koji Yamamoto, *Taming Capitalism before Its Triumph: Public Service, Distrust, and "Projecting" in Early Modern England* (Cambridge: Cambridge University Press, 2018).

84. Laura Brown, *Fables of Modernity: Literature and Culture in the English Eighteenth Century* (Ithaca, NY: Cornell University Press, 2001).

85. Carol Gibson, "Bernard Mandeville: The Importance of Women in the Development of Civil Societies" (MA thesis, University of British Columbia, 1989).

3. Enlightened Scarcity

1. Quoted in W. A. Speck, "Bernard Mandeville and the Middlesex Grand Jury," *American Society for Eighteenth-Century Studies* 11, no. 3 (1978), 363.

2. See Julian Hoppit, "The Myth of the South Sea Bubble," *Transactions of the Royal Historical Society* 12 (2002): 141–165.

3. Arnoud Orain, *La Politique du Merveilleux: Une Autre Histoire du Systèm de Law (1695–1795)* (Paris: Fayard, 2018).

4. Daniel Defoe, *The case of Mr. Law, truly stated. In answer to a pamphlet, entilul'd, A Letter to Mr. Law* (London, 1721).

5. Daniel Defoe, *An Essay upon Publick Credit.* (London, 1710), 51.

6. Jonathan Swift, "The South Sea Project," in *Poetical Works of Swift,* ed. Herbert Davis (London: Oxford University Press, 1967), 192–199.

7. Jonathan Swift, *Gulliver's Travels,* ed. Robert DeMaria Jr. (London: Benjamin Motte, Jr., 1726; London: Penguin, 2001), 166.

8. Swift, *Gulliver's Travels,* 168.

9. Swift, *Gulliver's Travels,* 172.

10. Caroline Merchant, *The Death of Nature: Women, Ecology, and the Scientific Revolution* (New York: HarperCollins, 1980).

11. Peter Reill, *Vitalizing Nature in the Enlightenment* (Chicago: University of Chicago Press, 2005), 7.

12. Quoted passage is authors' translation from a Swedish rendering of Linnaeus's 1752 dissertation (in Latin) on the usefulness of the natural sciences, "Cui Bono?" Carl Linnaeus, *En fråga, som altid föreställes de*

naturkunniga, då det heter: hwartil duger det? (Cui bono?) (Stockholm: L. Salvius, 1753), 3.

13. See, for example, Peter M. Jones, *Agricultural Enlightenment: Knowledge, Technology and Nature 1750–1840* (Oxford: Oxford University Press, 2016); Paola Bertucci, *Artisanal Enlightenment: Science and the Mechanical Arts in Old Regime France* (New Haven: Yale University Press, 2017).

14. Swift, *Gulliver's Travels,* 239.

15. Swift, *Gulliver's Travels,* 240.

16. Swift, *Gulliver's Travels,* 239.

17. Swift, *Gulliver's Travels,* 240.

18. Swift, *Gulliver's Travels,* 245.

19. Swift, *Gulliver's Travels,* 249.

20. Swift, *Gulliver's Travels,* 32.

21. Swift, *Gulliver's Travels,* 424.

22. Swift, *Gulliver's Travels,* 102.

23. Swift, *Gulliver's Travels,* 103.

24. Swift, *Gulliver's Travels,* 108.

25. Swift, *Gulliver's Travels,* 107.

26. Swift, *Gulliver's Travels,* 119.

27. Shaftesbury, *Characteristics of Men, Manners, Opinions, Times,* ed. Lawrence Klein (Cambridge: Cambridge University Press, 1999).

28. Francis Hutcheson, *An Inquiry into the Original of Our Ideas of Beauty and Virtue,* ed. Wolfgang Leidhold (1726; Indianapolis: Liberty Fund, 2004).

29. Joseph Butler, *Five Sermons,* ed. Stephen Darwell (Indianapolis: Hackett, 1983), 30.

30. Nicholas Xenos correctly points out that "Hume, as much as anyone, can lay claim to be the inventor of [modern] scarcity." *Scarcity and Modernity* (New York: Routledge, 1989), 21.

31. David Hume, *A Treatise of Human Nature,* ed. L. A. Selby-Bigge and P. H. Nidditch, 2nd ed. (1739; Oxford: Oxford University Press, 1978), 494.

32. David Hume, *Enquiries Concerning Human Understanding and Concerning the Principles of Morals,* ed. L. A. Selby-Bigge and P. H. Nidditch, 3rd ed. (1777; Oxford: Oxford University Press, 1975), 188.

33. Hume, *Enquiries,* 183.

34. Hume, *Enquiries,* 185.

35. Hume, *Enquiries,* 188.

36. David Hume, "Of Interest," in *Essays, Moral, Political, and Literary,* ed. Eugene Miller (Indianapolis: Liberty Fund, 1985), 300.

37. Maxine Berg, "In Pursuit of Luxury: Global History and British Consumer Goods in the Eighteenth Century," *Past & Present* 182, no. 1 (2004): 85–142.

38. Neil McKendrick, John Brewer, and J. H. Plumb, *The Birth of a Consumer Society: The Commercialization of Eighteenth-Century England*

(London: HarperCollins, 1982); Jan de Vries, *The Industrious Revolution: Consumer Behavior and the Household Economy, 1650 to the Present* (Cambridge: Cambridge University Press, 2008); and Roy Porter, *The Creation of the Modern World: The Untold Story of the British Enlightenment* (London: Norton, 2000).

39. David Hume, "The Sceptic," in *Essays, Moral, Political, and Literary,* ed. Eugene Miller (Indianapolis: Liberty Fund, 1985), 170–171.

40. David Hume, "Of Refinement in the Arts," in *Essays, Moral, Political, and Literary,* ed. Eugene Miller (Indianapolis: Liberty Fund, 1985), 270.

41. Hume, "Of Refinement," 271.

42. Hume, "Of Refinement," 270.

43. Hume, "Of Refinement," 271.

44. David Hume, "Of the Middle Station of Life," in *Essays, Moral, Political, and Literary,* ed. Eugene Miller (Indianapolis: Liberty Fund, 1985), 546.

45. David Hume, "Of the Delicacy of Taste and Passion," in *Essays, Moral, Political, and Literary,* ed. Eugene Miller (Indianapolis: Liberty Fund, 1985), 6–7.

46. Hume, "The Sceptic," 170.

47. Hume, "Of the Middle Station," 546–547.

48. Hume, "Of Refinement," 271. This argument is further elaborated in Margaret Schabas and Carl Wennerlind, *A Philosopher's Economist: Hume and the Rise of Capitalism* (Chicago: University of Chicago Press, 2020).

49. Hume, *Enquiries,* 283–284.

50. Adam Smith, *The Correspondence of Adam Smith,* ed. Ernest Campbell Mossner and Ian Simpson Ross (Oxford: Clarendon Press, 1987) 2; John Rae, *Life of Adam Smith* (London: Macmillan, 1895), 329.

51. S. Engler, F. Mauelshagen, J. Werner, and J. Luterbacher, "The Irish Famine of 1740–1741: Famine Vulnerability and 'Climate Migration,'" *Climate of the Past* 9 (2013): 1161–1179; Philipp Rössner, "The 1738–41 Harvest Crisis in Scotland," *Scottish Historical Review* 90, no. 1 (2011), 27–63; Adam Smith, *An Inquiry into the Nature and Causes of the Wealth of Nations,* 2 vols., ed. R. H. Campbell, A. S. Skinner, and W. B. Todd (Oxford: Oxford University Press, 1976), 1: 104.

52. Smith, *Wealth,* 1: 178.

53. Smith, *Wealth,* 1: 258.

54. Smith, *Wealth,* 1: 363.

55. Smith, *Wealth,* 1: 364.

56. Jessica Riskin, *Science in the Age of Sensibility: The Sentimental Empiricists of the French Enlightenment* (Chicago: University of Chicago Press, 2013).

57. The physiocrats believed that hoarding withdrew wealth from circulation and thus damaged the nation's capacity to generate affluence. Unlike Smith, they did not develop a theory of savings.

58. Mirabeau, *Philosophie Rurale,* quoted in Michael Kwass, "'Le Superflu, Chose Très Nécessaire': Physiocracy and Its Discontents in the Eighteenth-

Century Luxury Debate," in *The Economic Turn: Recasting Political Economy in Enlightenment Europe,* ed. Steven Kapland and Sophus Reinert (London: Anthem Press, 2019), 121.

59. Smith, *Wealth,* 1: 20–21, 405, 408–409; Fredrik Albritton Jonsson, *Enlightenment's Fable: The Scottish Highlands and the Origins of Environmentalism* (New Haven: Yale University Press, 2013), 129–130.

60. Adam Smith, *The Theory of Moral Sentiments,* ed. D. D. Raphael and A. L. Macfie (Oxford: Oxford University Press, 1976), 181–184.

61. Smith, *Theory,* 181–184.

62. Smith, *Theory,* 181–184.

63. Smith, *Theory,* 181–184; Lisa Hill, "'The Poor Man's Son' and the Corruption of Our Moral Sentiments: Commerce, Virtue and Happiness in Adam Smith," *Journal of Scottish Philosophy* 15, no. 1 (2017): 9–25.

64. Smith, *Theory,* 181–184.

65. Smith, *Theory,* 181–184.

66. Smith, *Theory,* 181–184.

67. Smith, *Wealth,* 1: 341.

68. Smith, *Wealth,* 1: 342.

69. Smith, *Wealth,* 1: 342.

70. Smith, *Wealth,* 1: 346.

71. Smith, *Wealth,* 1: 345.

72. Smith, *Wealth,* 1: 526, 539.

73. Smith, *Wealth,* 1: 88.

74. Smith, *Wealth,* 1: 625.

75. Smith, *Wealth,* 1: 111.

76. William Godwin, *An Enquiry Concerning Political Justice,* ed. Mark Philp (Oxford: Oxford University Press, 2013), 415.

77. Godwin, *Enquiry Concerning Political Justice,* 416.

78. Godwin, *Enquiry Concerning Political Justice,* 432.

79. Godwin, *Enquiry Concerning Political Justice,* 437.

80. Godwin, *Enquiry Concerning Political Justice,* 437.

81. Godwin, *Enquiry Concerning Political Justice,* 440.

82. Godwin, *Enquiry Concerning Political Justice,* 427.

83. Godwin, *Enquiry Concerning Political Justice,* 432.

84. Godwin, *Enquiry Concerning Political Justice,* 433.

85. Godwin, *Enquiry Concerning Political Justice,* 418.

4. Romantic Scarcity

1. Dorothy Wordsworth, *The Grasmere Journals,* ed. Pamela Woolf (Oxford: Oxford University Press, 1991), 112.

2. Dorothy Wordsworth, *Grasmere Journals,* 112. Lucy Newlyn, *William and Dorothy Wordsworth: All in Each Other* (Oxford: Oxford University Press,

2013), 237. For more on dwelling in nature, see Jonathan Bate, *The Song of the Earth* (London: Picador, 2000).

3. William Wordsworth, *The Major Works,* ed. Stephen Gill (Oxford: Oxford University Press, 2008), 177.

4. William Wordsworth, "Michael," in *Wordsworth's Poetry and Prose,* ed. Nicholas Halmi (New York: W. W. Norton, 2014), 147.

5. Dorothy Wordsworth, *The Grasmere and Alfoxden Journals* (Oxford: Oxford University Press, 2012), 112; Wordsworth, "Michael"; Newlyn, *William and Dorothy Wordsworth,* 138.

6. Jean-Jacques Rousseau, "Discourse on the Origin and the Foundations of Inequality among Men," in *The Discourses and Other Early Political Writings,* ed. Victor Gourevitch (Cambridge: Cambridge University Press, 1997), 167.

7. Rousseau, "Discourse on Inequality," 142–143.

8. Rousseau, "Discourse on Inequality," 162.

9. Rousseau, "Discourse on Inequality," 164.

10. Rousseau, "Discourse on Inequality," 164.

11. Rousseau, "Discourse on Inequality," 165.

12. Rousseau, "Discourse on Inequality," 166.

13. Rousseau, "Discourse on Inequality," 168.

14. Rousseau, "Discourse on Inequality," 177.

15. Rousseau, "Discourse on Inequality," 132. Rousseau freely admitted that the condition of the natural state was "conjectural" and "hypothetical" because the facts in the matter could never be fully known.

16. Jean-Jacques Rousseau, *Émile, or On Education,* ed. and trans. Allan Bloom (1762; New York: Basic Books, 1979), 97.

17. Rousseau, *Émile,* 213.

18. Rousseau, "Discourse on Inequality," 153.

19. Rousseau, *Émile,* 214.

20. Rousseau, "Discourse on Inequality," 171.

21. Pierre Force, *Self-Interest before Adam Smith: A Genealogy of Economic Science* (Cambridge: Cambridge University Press, 2008).

22. Rousseau, "Discourse on Inequality," 170.

23. Rousseau, *Émile,* 214.

24. Rousseau, "Discourse on Inequality," 171.

25. Rousseau, "Discourse on Inequality," 177.

26. Rousseau, "Last Reply [to Critics of The Discourse on the Sciences and Arts]," in *The Discourses and Other Early Political Writings,* ed. Victor Gourevitch (Cambridge: Cambridge University Press, 1997), 72.

27. Rousseau, "Last Reply," 73.

28. Rousseau, "Discourse on the Sciences and Arts," in *The Discourses and Other Early Political Writings,* ed. Victor Gourevitch (Cambridge: Cambridge University Press, 1997), 21.

29. Rousseau, "Last Reply," 73.

30. Rousseau, *Politics and the Arts: Letter to M. D'Alembert on the Theatre,* trans. Alan Bloom (Ithaca, NY: Cornell University Press, 1960), 60.

31. Rousseau, *Politics and the Arts,* 62.

32. Jean-Jacques Rousseau, "Considerations on the Government of Poland and on Its Planned Reformation," in Rousseau, *The Plan for Perpetual Peace, On the Government of Poland, and Other Writings on History and Politics,* ed. Christopher Kelly, trans. Christopher Kelly and Judith Bush (Hanover, NH: Dartmouth College Press, 2011), 2:183.

33. Rousseau, "Considerations," 126.

34. Rousseau, *Émile,* 47.

35. Rousseau, *Émile,* 188, 195.

36. Rousseau, *Émile,* 195.

37. On this fatal contradiction, see Eoin Daly, "Providence and Contingency in Corsica: Rousseau on Freedom without Politics," *European Journal of Political Theory* 20, no. 4 (2021): 739–760, 755.

38. Rousseau, *Reveries of a Solitary Walker,* trans. Russell Goulbourne (Oxford: Oxford University Press, 2011), 56.

39. Rousseau, *Reveries,* 55.

40. Rousseau, "Discourse on Inequality," 143. Neuchâtel was affiliated with the Swiss confederation but also a Prussian principality (1707–1848).

41. Robert Darnton, *The Great Cat Massacre and Other Episodes in French Cultural History* (New York: Basic Books, 1984), 231.

42. Jacques-Henri Bernardin de Saint-Pierre, *Paul and Virginia* (Boston: Houghton, Mifflin, 1867), 33.

43. For the influence of Bernardin de Saint-Pierre on the Wordsworths, see Elizabeth Fay, *Becoming Wordsworthian: A Performative Aesthetics* (Amherst: University of Massachusetts Press, 1995), 54; Dorothy Wordsworth, *Journals of Dorothy Wordsworth,* ed. William Knight, 2 vols. (London: MacMillan, 1897), 1: viii.

44. Daniel Maudlin, *The Idea of the Cottage in English Architecture, 1760–1860* (London: Routledge, 2015), 3–5, 93–94.

45. *The Early Letters of William and Dorothy Wordsworth,* ed. Ernest de Selincourt (Oxford: Clarendon Press, 1935), 94.

46. Wordsworth, "Michael," 147.

47. Annabel Patterson, *Pastoral and Ideology: Virgil to Valéry* (Berkeley: University of California Press, 1988), 274, 279–280.

48. Wordsworth, "Michael," 148.

49. Wordsworth, "Michael," 152.

50. Wordsworth, "Michael," 148.

51. Wordsworth, "Michael."

52. Terry McCormick, "Wordsworth and Shepherds," in *The Oxford Handbook of William Wordsworth,* ed. Richard Gravil and Daniel Robinson (Oxford:

Oxford University Press, 2015), 643–644; Hardwicke Rawnsley, *Ruskin and the English Lakes* (Glasgow: J. MacLehose and Sons, 1901), 169.

53. John Clare, "Helpstone," in *Major Works,* ed. Eric Robinson and David Powell (Oxford: Oxford University Press, 2008), 4.

54. Clare, "Remembrances," in *Major Works,* 258–259.

55. Clare, "Remembrances," 260.

56. Clare, "The Lamentations of Round-Oak Waters," in *Major Works,* 21.

57. Clare, "Helpstone," 4.

58. John Law Cherry, *Life and Remains of John Clare: The Northamptonshire Peasant Poet,* (London: Frederick Warne, 1873), 109.

59. J. M. Neeson, *Commoners: Common Right, Enclosure and Social Change in England 1700–1820* (Cambridge: Cambridge University Press, 1993), 40–41.

60. Clare, "Remembrances," 260.

61. Clare, "The Robins Nest," in *Major Works,* 223–224.

62. Clare, "To the Snipe," in *Major Works,* 206.

63. John Stuart Mill, *Collected Works,* vol. 32: *Additional Letters of John Stuart Mill,* ed. Marion Filipiuk et al. (Toronto: University of Toronto Press, 1991), 146–147.

64. Mill, *Collected Works,* 32: 146–147; *Proceedings of the Royal Horticultural Society* 4, 91–93, quote on 91.

65. John Stuart Mill, *The Principles of Political Economy,* 2 vols. (London: John W. Parker, 1848), 2: 306.

66. Mill, *Principles,* 2: 312.

67. Mill, *Principles,* 2: 308.

68. Mill, *Principles,* 2: 308.

69. Mill, *Principles,* 2: 311–312.

70. Mill, *Principles,* 2: 311.

71. Mill, *Principles,* 2: 119.

72. Mill, *Principles,* 1: 332.

73. Mill, *Principles,* 1: 300; see also Donald Winch, *Wealth and Life: Essays on the Intellectual History of Political Economy in Britain, 1848–1914* (Cambridge: Cambridge University Press, 2009), 61–67.

74. Tim Hilton, *John Ruskin* (New Haven: Yale University Press, 2002), 494, 497, 505; Sarah Haslam, *John Ruskin and the Lakeland Arts Revival, 1880–1920* (Cardiff: Merton Priory Press, 2004). This analysis of Ruskin draws from arguments in Vicky Albritton and Fredrik Albritton Jonsson, *Green Victorians: The Simple Life in John Ruskin's Lake District* (Chicago: University of Chicago Press, 2016); and Fredrik Albritton Jonsson, "Political Economy," in *Historicism and the Human Sciences in Victorian Britain,* ed. Mark Bevir (Cambridge: Cambridge University Press, 2017).

75. John Ruskin, *Works of John Ruskin,* vol. 27: *Letters 1–36. Fors clavigera,* ed. E. T. Cook and Alexander Wedderburn (London: George Allen, 1907), 527.

76. Winch, *Wealth and Life,* 63–68; William Stanley Jevons, *The Coal Question: An Inquiry Concerning the Progress of the Nation, and the Probable Exhaustion of Our Coal-mines,* 3rd ed., rev. (New York: Augustus M. Kelley, 1965), 15–18.

77. Ruskin, *Works,* 27: 91.

78. John Ruskin, "Ad Valorem," in *Works of John Ruskin,* vol. 17: *Unto this Last, Muneral Pulvis, Time and Tide with Other Writings on Political Economy 1860–1873,* ed. E. T. Cook and Alexander Wedderburn (London: George Allen, 1905), 94.

79. Ruskin, "Ad Valorem," 113.

80. John Ruskin, "Wealth," in *Works of John Ruskin,* 17: 153.

5. Malthusian Scarcity

1. Samuel Taylor Coleridge, *Collected Letters of Samuel Taylor Coleridge, Volume IV: 1815–1819,* ed. Earl Leslie Griggs (Oxford: Oxford University Press, 2002), 660. Quoted in Gillen D'Arcy Wood, *Tambora: The Eruption that Changed the World* (Princeton: Princeton University Press, 2014) 49, 51, 55–58, 67, 70.

2. Wood, *Tambora,* 61; David Ricardo, *The Works and Correspondence of David Ricardo,* vol. 7: *Letters 1816–18,* ed. Piero Sraffa (Cambridge: Cambridge University Press, 1952), 62, 66, 68; Mill quoted in Amartya Sen, *The Idea of Justice* (Cambridge, MA: Belknap Press of Harvard University Press, 2009), 388.

3. Roger A. E. Wells, *Wretched Faces: Famine in Wartime England 1793–1801* (New York: St. Martin's Press, 1988), 40–41.

4. Wells, *Wretched Faces,* 4, 110–119, quotation on 119.

5. E. P. Thompson, *Customs in Common: Studies in Traditional Popular Culture* (New York: New Press, 1993). See especially chap. 4, "The Moral Economy of the English Crowd in the Eighteenth Century" and chap. 5, "The Moral Economy Reviewed."

6. Donald Winch, *Riches and Poverty: An Intellectual History of Political Economy in Britain, 1750–1834* (Cambridge: Cambridge University Press, 1996), 200–201.

7. Edmund Burke, *Thoughts and Details on Scarcity* (London: F. and C. Rivington, 1800), 23.

8. Burke, *Thoughts,* 25.

9. Burke, *Thoughts,* 25.

10. Burke, *Thoughts,* 26.

11. Burke, *Thoughts,* 4, 32.

12. Burke, *Thoughts,* 32.

13. Burke, *Thoughts,* 18.

14. Burke, *Thoughts,* 18.

15. Emma Rothschild, *Economic Sentiments: Adam Smith, Condorcet, and the Enlightenment* (Cambridge, MA: Harvard University Press, 2013), 263; Edmund Burke, *Reflections on the Revolution in France* (Oxford: Oxford University Press, 1993), 90, 99.

16. Burke, *Thoughts,* 32.

17. Burke, *Thoughts,* 34.

18. Burke, *Thoughts,* 35.

19. Burke, *Thoughts,* 35.

20. Burke, *Thoughts,* 44.

21. Burke, *Thoughts,* 44.

22. Burke, *Thoughts,* 44.

23. Burke, *Thoughts,* 31.

24. Burke, *Thoughts,* 47.

25. Anon. [Thomas R. Malthus], *An Essay on the Principle of Population, As It Affects the Future Improvement of Society* (London: J. Johnson, 1798), 14.

26. Malthus, *Essay* (1798), 48.

27. Malthus, *Essay* (1798), 15.

28. Malthus, *Essay* (1798), 26.

29. Malthus, *Essay* (1798), 29.

30. Thomas R. Malthus, *An Essay on the Principle of Population . . . A New Edition, very much Enlarged,* 2nd ed. (London: J. Johnson, 1803), 75.

31. Malthus, *Essay* (1803), 32, 548.

32. Compare Malthus, *Essay* (1803), 17, 58, 95, 324–329. See also his speculation about "a standard of wretchedness" among the lower classes "in most countries," 557.

33. Malthus, *Essay* (1798), 139.

34. Malthus, *Essay* (1798), 48.

35. Malthus, *Essay* (1803), 429.

36. Malthus, *Essay* (1803), 427.

37. Fredrik Albritton Jonsson, "Island, Nation, Planet: Malthus in the Enlightenment," in *New Perspectives on Malthus,* ed. Robert Mayhew (Oxford: Oxford University Press, 2016), 128–154.

38. Malthus, *Essay* (1803), 464.

39. Malthus, *Essay* (1803), 464.

40. Malthus, *Essay* (1803), 526.

41. Malthus, *Essay* (1803), 525–527.

42. Malthus, *Essay,* 5th ed., 3 vols. (1817), 2: 501, 3: 145. Malthus softened his rhetoric over time but did not abandon his pragmatic defense of protectionism in the name of national security. Compare John Pullen, "Malthus on Agricultural Protection: An Alternative View," *History of Political Economy* 27, no. 3 (1995): 517–529.

43. David Ricardo, *The Principles of Political Economy and Taxation* (London, 1817), 158–159.

44. Ricardo, *Principles of Political Economy*, 160n..

45. Ricardo, *The Principles of Political Economy and Taxation,* 3rd ed. (London, 1821), 313n.

46. Ricardo, *Principles of Political Economy,* 3rd ed.,, 313n.

47. Ricardo, *Principles of Political Economy,* 3rd ed., 154.

48. Ricardo, *Principles of Political Economy,* 3rd ed., 115.

49. Catherine Heyrendt-Sherman, "Gender Revolution in a Malthusian Utopia: Harriet Martineau's World of Garveloch," *Cahiers victorien et édouardien* 89 (2019): 1–14.

50. Malthus, *Essay* (1803), 553 note b.

51. For Victoria's interest in "Ella of Garveloch," see Harriet Martineau, *Harriet Martineau's Autobiography,* ed. Maria Weston Chapman, 2 vols. (Boston: Houghton Osgood, 1877), 2: 209.

52. Harriet Martineau, "Ella of Garveloch," in *Illustrations of Political Economy No. 5,* 2nd ed. (London: C. Fox, 1832), 126–127.

53. Harriet Martineau, "Weal and Woe in Garveloch," in *Illustrations of Political Economy No. 6,* 2nd ed. (London: C. Fox, 1832), 69.

54. Martineau, "Weal and Woe," 68.

55. Martineau, "Weal and Woe," 64–65, 68.

56. Martineau, "Weal and Woe," 74.

57. Martineau, "Weal and Woe," 135.

58. For the legacy of this strand of Malthusianism, see Alison Bashford, *Global Population: History, Geopolitics, and Life on Earth* (New York: Columbia University Press, 2014), chap. 8, 211–238.

59. On the hidden hand, see Boyd Hilton, "The Role of Providence in Evangelical Social Thought," in *History, Society and the Churches: Essays in Honour of Owen Chadwick,* ed. Derek Beales and Geoffrey Best, 215–234 (Cambridge: Cambridge University Press, 1985), 233.

60. Charles Edward Trevelyan to Thomas Spring-Rice, Lord Monteagle, October 9, 1846, Monteagle Papers, National Library of Ireland, Dublin, Ms. 13, 397 / 1. The letter is reproduced with some editing in Noel Kissane, *The Irish Famine: A Documentary History* (Dublin: National Library of Dublin, 1995), 51.

61. C. E. Trevelyan, *The Irish Crisis* (London: Longman, Brown, 1848), 4–5, 196.

62. Charles Darwin, *On the Origin of Species,* ed. with intro. and notes by Gillian Beer (1859; Oxford: Oxford University Press, 2008), 65.

63. Charles Darwin, *The Formation of Vegetable Mould, Through the Action of Worms, with Observations on Their Habits* (London: John Murray, 1881), 313.

64. Darwin, *Origin of Species,* 66.

65. Darwin, *Origin of Species,* 360.

66. Peter Kropotkin, *Mutual Aid: A Factor of Evolution* (London: Freedom Press, 2009).

67. Kropotkin's perspective proved generative of subsequent approaches to the question of competition versus cooperation in the natural worlds. See, for example, Stephen Jay Gould, "Kropotkin Was No Crackpot," *Natural History* 97, no. 7 (1988): 12–21; and Nicholas Christakis, *Blueprint: The Evolutionary Origins of a Good Society* (New York: Little, Brown, 2019).

68. Ruth Kinna, "Kropotkin and Huxley," *Politics* 12, no. 2 (1992): 41–47.

69. Martin Daunton, "Society and Economic Life," in *The Nineteenth Century: British Isles, 1815–1901,* ed. Colin Matthew (Oxford: Oxford University Press, 2000), 42–43; Bashford, *Global Population.*

70. Frank Trentmann, *Free Trade Nation: Commerce, Consumption and Civil Society in Modern Britain* (Oxford: Oxford University Press, 2009); James Belich, *Replenishing the Earth: The Settler Revolution and the Rise of the Angloworld* (Oxford: Oxford University Press, 2009); Chris Otter, *Diet for a Large Planet: Industrial Britain, Food Systems, and World Ecology* (Chicago: University of Chicago Press, 2020), 27.

71. Will Steffen, Paul J. Crutzen, and John R. McNeill, "The Anthropocene: Are Humans Now Overwhelming the Great Forces of Nature?" *Ambio* 36, no. 8 (2007): 614–621, Table 1.

6. Socialist Scarcity

1. Albert Tangeman Volwiler, "Robert Owen and the Congress of Aix-la-Chapelle, 1818," *Scottish Historical Review* 19, no. 74 (1922): 96–105.

2. E. A. Wrigley, *Energy and the English Industrial Revolution* (Cambridge: Cambridge University Press, 2010); James Belich, *Replenishing the Earth: The Settler Revolution and the Rise of the Angloworld* (Oxford: Oxford University Press, 2011).

3. William Cavert, *The Smoke of London: Energy and Environment in the Early Modern City* (Cambridge: Cambridge University Press, 2016).

4. William Ashworth, *The Industrial Revolution: The State, Knowledge and Global Trade* (London: Bloomsbury Academic, 2017).

5. Will Steffen, Paul J. Crutzen, and John R. McNeill, "The Anthropocene: Are Humans Now Overwhelming the Great Forces of Nature?" *Ambio* 36, no. 8 (2007): 614–621, 617, Table 1.

6. Friedrich Engels, *The Condition of the Working Class in England* (Oxford: Oxford University Press, 2009), 79, 84.

7. Engels, *Condition of the Working Class,* 109. For a summary of the literature on urban mortality, and a revisionist critique, see Romola Davenport, "Urbanization and Mortality, c. 1800–1850," *Economic History Review* 73, no. 2 (2020): 455–485.

8. Robert Owen, *A Supplementary Appendix to the First Volume of the Life of Robert Owen,* vol. 1A (London: Effingham Wilson, 1858), 215.

9. Owen, *A Supplementary Appendix,* 216.

10. Robert Owen, *A New View of Society: Or, Essays on the Formation of the Human Character,* 2nd ed. (London: Longman, 1816), 175.

11. Owen, *New View of Society,* 175.

12. Owen, *Supplementary Appendix,* 215.

13. Owen, *Supplementary Appendix,* 211.

14. Owen, *Supplementary Appendix,* 215.

15. Owen, *Supplementary Appendix,* 215.

16. Quoted in Jonathan Beecher, *Charles Fourier: The Visionary and His World* (Berkeley: University of California Press, 1986), 369.

17. Beecher, *Charles Fourier,* 87, 26

18. Owen praised Fourier's critique: "I must thank you personally, sir, for the pleasure I derived from the reading of your work. Your tableaux of the vices of civilization are charming in their truth and in their strength." Quoted in Beecher, *Charles Fourier,* 368.

19. Keith Taylor, *The Political Ideas of the Utopian Socialists* (London: Frank Cass, 1982), 108–110.

20. In exploring the links between the material world, the realm of the passions, and the celestial sphere, Fourier embraced lessons from mysticism, perhaps from Emmanuel Swedenborg or the earlier alchemical thinkers. Beecher, *Charles Fourier,* 341–342.

21. Charles Fourier, *The Theory of the Four Movements,* ed. Gareth Stedman Jones and Ian Patterson (Cambridge: Cambridge University Press, 1996), 251.

22. With people spending only a fraction of their day in manufacturing, Fourier's vision of the future was a world of peaceful, bucolic, agricultural bliss.

23. Charles Fourier, *The Hierarchies of Cuckoldry and Bankruptcy,* trans. Geoffrey Longnecker (Cambridge, MA: Wakefield Press, 2011).

24. Quoted in Beecher, *Charles Fourier,* 21.

25. Fourier, *Theory of the Four Movements,* 50.

26. Beecher, *Charles Fourier,* 223.

27. Karl Marx, "Theses on Feuerbach" (1845), in *Karl Marx: Selected Writings,* ed. David McLellan (Oxford: Oxford University Press, 1977), 158.

28. Karl Marx and Friedrich Engels, *The Communist Manifesto,* trans. Samuel Moore (Harmondsworth: Penguin, 1967), 97.

29. Karl Marx, *The Economic and Philosophical Manuscripts of 1844,* ed. Dirk Struik (New York: International Publishers, 1964), 116.

30. Karl Marx, *Grundrisse: Foundations of the Critique of Political Economy,* trans. Martin Nicolaus (London: Penguin, 1973), 713–714.

31. Marx uses the term *alienation* a handful of times in *Capital.* For example, in a passage that is clearly drawn from his *Economic and Philosophic Manuscripts of 1844,* he writes that, in capitalism, "all means for the development of production undergo a dialectical inversion so that they become means of domination and exploitation of the producers; they distort the worker into a

fragment of a man, they degrade him to the level of an appendage of a machine, they destroy the actual content of his labour by turning it into a torment; they alienate from him the intellectual potentialities of the labour process in the same proportion as science is incorporated in it as an independent power." Karl Marx, *Capital,* vol. 3 (Harmondsworth: Penguin, 1991), 799.

32. Karl Marx, *A Contribution to the Critique of Political Economy,* ed. Maurice Dobb, trans. S. W. Ryazanskaya (Moscow: Progress Publishers, 1970), 30. See also Harry Cleaver, *Reading* Capital *Politically* (Austin: University of Texas Press, 1979), 109.

33. Marx, *Capital,* 230.

34. Marx, *Capital,* 253.

35. Marx, *Capital,* 255.

36. The capitalist, in his "vampire-like" quest, "only lives by sucking living labor, and lives the more, the more labor it sucks." Marx, *Capital,* 254.

37. Karl Marx to Pavel Annenkov, December 28, 1846, *Letters of Karl Marx,* ed. Saul Padover (Englewood Cliffs, NJ: Prentice-Hall, 1979), 48. Marx incorporated this line some twenty years later in *Capital,* noting that "It would be possible to write a whole history of the inventions made since 1830 for the sole purpose of providing capital with weapons against working-class revolt." Marx, *Capital,* 563.

38. Marx, *Capital,* 909.

39. Marx, *Capital,* 617.

40. Marx, *Capital,* 693.

41. Karl Marx, *Grundrisse: Foundations of the Critique of Political Economy (Rough Draft),* trans. Martin Nicolaus (London: Penguin, 1973), 693.

42. Marx and Engels, *Communist Manifesto,* 87.

43. Karl Marx to Arnold Ruge, Kreuznach, September 1843, in "Letters from the *Deutsch-Französische Jahrbücher,*" in *Collected Works,* vol. 3: *Marx and Engels: 1843–1844* (London: Lawrence and Wishart, 1975), 142.

44. Karl Marx and Friedrich Engels, *The German Ideology* (1846), in *Karl Marx Selected Writings,* 169.

45. Marx and Engels, *Communist Manifesto,* 99.

46. Marx and Engels, *Communist Manifesto,* 83.

47. Marx, *Economic and Philosophic Manuscripts of 1844,* 135.

48. Marx, *Capital,* 506–507.

49. Marx, *Capital,* 503.

50. Marx, *Capital,* 949.

51. Marx, *Economic and Philosophic Manuscripts of 1844,* 135.

52. Marx, *Capital,* 134.

53. Marx, *Capital,* 289.

54. Marx, *Capital,* 637.

55. Marx, *Capital,* 637.

56. Marx, *Capital,* 784. Repudiating Malthus, Marx offered up his own law of population rooted in the organic composition of capital rather than natural limits of soil and population.

57. Marx, *Capital,* 766n6.

58. Kohei Saito, *Karl Marx's Ecosocialism: Capital, Nature and the Unfinished Critique of Political Economy* (New York: Monthly Review Press, 2017).

59. Rosa Luxemburg, *The Mass Strike,* ed. Tony Cliff (1906; London: Bookmarks, 1986), 46.

60. Vladimir Lenin, "The Urgent Problems of the Soviet Rule," *Pravda,* March 26, 1918, in D. Del Mar and R. D. Collons, eds., *Classics in Scientific Management* (Tuscaloosa: University of Alabama Press, 1976), 377.

61. Lenin, "The Urgent Problems of the Soviet Rule," 377.

62. Lenin, "The Urgent Problems of the Soviet Rule," 377.

63. Gregg Marland, Thomas A. Boden, Robert J. Andres, *Global, Regional, and National CO$_2$ Emissions in Trends: A Compendium of Data on Global Change* (Oak Ridge, TN: Oak Ridge National Laboratory, US Department of Energy, 2000).

7. Neoclassical Scarcity

1. Karl Marx and Friedrich Engels, *The Communist Manifesto,* trans. Samuel Moore (Harmondsworth: Penguin, 1967), 78.

2. Vaclav Smil, *Creating the Twentieth Century: Technical Innovations of 1867–1914 and Their Lasting Impact* (Oxford: Oxford University Press, 2005).

3. Francesco Boldizzoni, *Foretelling the End of Capitalism: Intellectual Misadventures since Karl Marx* (Cambridge, MA: Harvard University Press, 2020).

4. William Morris, *How I Became a Socialist* (1894; London: Twentieth Century Press, 1896), 9.

5. Thomas Hodgskin, *Labour Defended against the Claims of Capital* (London: Knight and Lacey, 1825), 27.

6. Knud Wicksell, *Lectures in Political Economy* (1901; London: Routledge, 1934), 28.

7. See Margaret Schabas, *A World Ruled by Number: William Stanley Jevons and the Rise of Mathematical Economics* (Princeton: Princeton University Press, 1990); Harro Maas, *William Stanley Jevons and the Making of Modern Economics* (Cambridge: Cambridge University Press, 2005).

8. Jevons, *Pure Logic and Other Minor Works* (London: Macmillan, 1890), 201.

9. Jevons, "The Importance of Diffusing a Knowledge of Political Economy," lecture, Owens College, Manchester, UK, October 12, 1866.

10. Deborah Coen, *Vienna in the Age of Uncertainty: Science, Liberalism, and Private Life* (Chicago: University of Chicago Press, 2007).

11. Léon Walras quoted in Keith Tribe, *The Economy of the Word: Language, History, and Economics* (Oxford: Oxford University Press, 2015), 276.

12. Carl Menger, *Principles of Economics,* trans. J. Dingwall and B. Hoselitz (1871; Grove City, PA: Libertarian Press, 1994), 120–121.

13. William Stanley Jevons, *The Theory of Political Economy* (1871; New York: Kelley and Millman, 1957), 1.

14. The concept of diminishing marginal utility helped settle the diamond-water paradox—that is, the question of why relatively unnecessary commodities, such as diamonds, can be extremely expensive while substances essential to life, such as water, can be free. Diminishing marginal utility also explains why demand curves slope downward.

15. Jevons, *Theory of Political Economy,* vi.

16. Walras suggested not that his general equilibrium theory constituted an accurate representation of current conditions but that it served as an ideal toward which a society should strive.

17. Menger, *Principles of Economics,* 77; Jevons, *Theory of Political Economy,* 304.

18. Léon Walras, *Elements of Pure Economics, or the Theory of Social Wealth,* trans. William Jaffé (1874; Homewood, IL: Richard Irwin, 1954), 65. Also see Philip Mirowski, *More Heat Than Light: Economics as Social Physics, Physics as Nature's Economics* (Cambridge: Cambridge University Press, 1989).

19. Menger, *Principles of Economics,* 75.

20. Menger, *Principles of Economics,* 74.

21. Jevons, *Theory of Political Economy,* 40 (emphasis added). Menger similarly argued that it was an inevitable evolution, part of the basic dynamics of human progress, that "the satisfaction of every lower want in the scale creates a desire of a higher character."

22. Walras, *Elements of Pure Economics,* 66–67.

23. Thorstein Veblen, *The Theory of the Leisure Class* (1899; New York: Dover, 1994), 31.

24. Veblen, *Theory of the Leisure Class,* 47.

25. Joris-Karl Huysmans, *Against Nature: A New Translation by Robert Baldick* (1884; London: Penguin, 1959).

26. Huysmans, *Against Nature,* 54.

27. Huysmans, *Against Nature,* 55.

28. Quoted in Lisa Tiersten, *Marianne and the Market: Envisioning Consumer Society in Fin-de-Siècle France* (Berkeley: University of California Press, 2001), 15.

29. Tiersten, *Marianne and the Market,* chap. 1.

30. Walras, *Elements of Pure Economics,* 65.

31. Rosalind Williams, *Dream Worlds: Mass Consumption in Late Nineteenth-Century France* (Berkeley: University of California Press, 1982), 288.

32. Simon Clarke, *Marx, Marginalism, and Modern Sociology: From Adam Smith to Max Weber* (London: Macmillan, 1982); Julie Matthaei, "Rethinking Scarcity: Neoclassicism, NeoMalthusianism, and NeoMarxism," *Review of Radical Political Economics* 16, no. 2/3 (1984): 81–94; Nicholas Xenos, *Scarcity and Modernity* (London: Routledge, 1989); Carl Wennerlind, "The Historical Specificity of Scarcity" (PhD diss., University of Texas at Austin, 1999); Costas Panayotakis, *Remaking Scarcity: From Capitalist Inefficiency to Economic Democracy* (London: Pluto Press, 2011).

33. Other prominent second-generation marginalists included Vilfredo Pareto, Eugen von Böhm-Bawerk, and Francis Ysidro Edgeworth, all of whom made significant contributions and helped further popularize the new economics.

34. Alfred Marshall, "Some Features of American Industry," in *The Early Economic Writings of Alfred Marshall, 1867–1890,* vol. 2, ed. John Whitaker (London: Royal Economic Society, 1975), 354.

35. Alfred Marshall, "The Social Possibilities of Economic Chivalry," *Economic Journal* 17, no. 65 (1907): 7–29, 9.

36. Alfred Marshall, *Principles of Economics,* ed. C. W. Guillebaud (1890; New York: Macmillan, 1961), 86.

37. Marshall, *Principles of Economics,* 87.

38. Marshall, *Principles of Economics,* 3.

39. Alfred Marshall, "The Future of the Working Classes," in *Memorials of Alfred Marshall,* ed. A. C. Pigou (London: Macmillan, 1925), 101. Originally read at the Conversazione of the Cambridge "Reform Club" on November 25, 1873, and published in *The Eagle,* the magazine of St John's College, Cambridge University.

40. Marshall, "The Future of the Working Classes," 10.

41. Marshall, "The Future of the Working Classes," 14. Marshall points out that an essential feature of this progress was that the working classes continued the trend of having fewer children. A larger population would create Malthusian pressures and make it more difficult to maintain high standards of education for the masses, which would eliminate any progress achieved.

42. Marshall, "The Social Possibilities," 9.

43. Marshall, "The Social Possibilities," 9.

44. Marshall, "The Social Possibilities," 25.

45. John Maynard Keynes, "Economic Possibilities for Our Grandchildren," in *Essays in Persuasion* (1930; New York: Norton, 1963), 365.

46. Keynes, "Economic Possibilities for Our Grandchildren," 365.

47. Keynes, "Economic Possibilities for Our Grandchildren," 367.

48. John Stuart Mill, *Principles of Political Economy,* 751.

49. Keynes, "Economic Possibilities for Our Grandchildren," 366.

50. Keynes, "Economic Possibilities for Our Grandchildren," 368.

51. Keynes, "Economic Possibilities for Our Grandchildren," 372.

52. Stephen Marglin, *The Dismal Science: How Thinking Like an Economist Undermines Community* (Cambridge, MA: Harvard University Press, 2010), 201.

53. Jevons, "The Future of Political Economy, Introductory Lecture at the Opening of the Session 1876-7, at University College, London, Faculty of Arts and Laws," *Fortnightly Review* 20, no. 129 (1876), 623.

54. Kenneth Arrow and Gerard Debreu, "Existence of an Equilibrium for a Competitive Economy," *Econometrica* 22, no. 3 (1954): 265-290.

55. See Till Düppe and Roy Weintraub, *Finding Equilibrium: Arrow, Debreu, McKenzie and the Problem of Scientific Credit* (Princeton: Princeton University Press, 2014), 136.

56. Philip Mirowski argues that neoclassical economics served as a battle ram during the Cold War, whereas Düppe and Weintraub insist that the main inspiration behind economics at the RAND Corporation was the mathematization of the discipline. See Philip Mirowski, *Machine Dreams: Economics Becomes a Cyborg Science* (Cambridge: Cambridge University Press, 2002); and Düppe and Weintraub, *Finding Equilibrium*.

57. Michel Foucault, *The Birth of Biopolitics: Lectures at the Collège de France, 1978–1979*, ed. Michel Senellart, trans. Graham Burchell (London: Macmillan, 2010).

58. Quoted in Paul Sabin, *The Bet: Paul Ehrlich, Julian Simon, and Our Gamble over Earth's Future* (New Haven: Yale University Press, 2013), 147.

59. Arthur C. Pigou, *The Economics of Welfare* (London: Macmillan, 1960).

60. Ian Kumekawa, *The First Serious Optimist: A. C. Pigou and the Birth of Welfare Economics* (Princeton: Princeton University Press, 2017).

61. Steven Medema, *The Hesitant Hand: Taming Self-Interest in the History of Economic Ideas* (Princeton: Princeton University Press, 2009), 106.

62. Ronald Coase, "The Problem of Social Cost," *Journal of Law & Economics* 3 (1960): 1–44.

63. Ronald Coase, "Social Cost and Public Policy," in *Exploring the Frontiers of Administration,* ed. George Edwards (Toronto: York University Press, 1970), 43.

64. This framework gave rise to cap-and-trade, which is discussed in the Conclusion.

65. J. H. Dales, *Pollution, Property and Prices: An Essay in Policy-Making and Economics* (Toronto: University of Toronto Press, 1968); and David Montgomery, "Markets in Licenses and Efficient Pollution Control Programs," *Journal of Economic Theory* 5, no. 3 (1972): 395–418.

66. Alfred Marshall, in *The Correspondence of Alfred Marshall, Economist,* vol. 2, ed. John Whitaker (Cambridge: Cambridge University Press, 1996), 127.

67. Marshall in *The Correspondence of Alfred Marshall,* 127.

68. Quoted in Thomas Robertson, "Development," in *Encyclopedia of the Cold War,* vol. 1, ed. Ruud van Dijk (London: Routledge, 2008), 255.

69. Matthias Schmelzer, *The Hegemony of Growth: The OECD and the Making of the Economic Growth Paradigm* (Cambridge: Cambridge University Press, 2016).

70. Robert Solow, "The Economics of Resources or the Resources of Economics," *American Economic Review* 64, no. 2 (1974): 1–14, 11.

71. William Nordhaus and James Tobin, "Is Growth Obsolete?" in *Economic Research: Retrospect and Prospect,* vol. 5: *Economic Growth* (New York: NBER, 1972), 522.

72. Nordhaus and Tobin, "Is Growth Obsolete?" 522.

73. William Nordhaus, *A Question of Balance: Weighing the Options on Global Warming Policies* (New Haven: Yale University Press, 2008).

74. Christopher Jones, *The Invention of Infinite Growth* (Chicago: University of Chicago Press, *forthcoming*).

75. While neoclassical economics was not yet dominant in Europe by the early 1970s, by the end of the following decade most major European universities had embraced it. The creation of the Nobel Prize in Economics played a significant role in this process. See Avner Offer and Gabriel Söderberg, *The Nobel Factor: The Prize in Economics, Social Democracy, and the Market Turn* (Princeton: Princeton University Press, 2016).

76. Roger Backhouse, *Founder of Modern Economics: Paul A. Samuelson.* vol. 1: *Becoming Samuelson, 1915–48* (Oxford: Oxford University Press, 2017).

77. While scarcity was the centerpiece of neoclassical economics from the marginalists onward, it was only by the early 1960s that textbooks universally defined economics as the science of choice under the condition of scarcity. Roger Backhouse and Steven Medema, "On the Definition of Economics," *Journal of Economic Perspectives* 23, no. 1 (2009): 221–233.

78. Lord Robbins, *An Essay on the Nature and Significance of Economic Science* (1932; New York: New York University Press, 1981), 15.

79. Francis Fukuyama, *The End of History and the Last Man* (New York: Free Press, 1992).

80. Marglin, *The Dismal Science,* 216.

81. Ronald Reagan, Remarks at Convocation Ceremonies at the University of South Carolina in Columbia, September 20, 1983, https://www.presidency.ucsb.edu/documents/remarks-convocation-ceremonies-the-university-south-carolina-columbia.

8. Planetary Scarcity

1. Kendrick Oliver, *To Touch the Face of God* (Baltimore: Johns Hopkins University Press, 2013), 145–146.

2. Lovell and Anders quoted in Oliver, *To Touch the Face of God,* 103, 145.

3. Tim Lenton, *Earth System Science: A Very Short Introduction* (Oxford: Oxford University Press, 2016).

4. John R. McNeill and Peter Engelke, *The Great Acceleration: An Environmental History of the Anthropocene since 1945* (Cambridge, MA: Belknap Press of Harvard University Press, 2014).

5. McNeill and Engelke, *Great Acceleration,* 10; Will Steffen, Paul Crutzen, and John McNeill, "The Anthropocene: Are Humans Now Overwhelming the Great Forces of Nature?" *Ambio* 36, no. 8 (2007): 614–621, 617, Table 1; Johan Rockström et al., "Planetary Boundaries: Exploring the Safe Operating Space for Humanity," *Ecology and Society* 14, no. 2 (2009).

6. Robert Collins, *More: The Politics of Growth in Postwar America* (Oxford: Oxford University Press, 2002); Stephen Macekura, *The Mismeasure of Progress: Economic Growth and Its Critics* (Chicago: University of Chicago Press, 2020).

7. Our concept of Planetary Scarcity draws crucial inspiration from Dipesh Chakrabarty's seminal concept of the Planetary Age. Dipesh Chakrabarty, *The Climate of History in a Planetary Age* (Chicago: University of Chicago Press, 2021).

8. Joshua Howe, *Behind the Curve: Science and Politics of Global Warming* (Seattle: University of Washington Press, 2016); Paul Edwards, *A Vast Machine: Computer Models, Climate Data and the Politics of Global Warming* (Cambridge, MA: MIT Press, 2013).

9. Hannah Arendt, *The Human Condition,* 2nd ed. (1958; Chicago: University of Chicago Press, 1998), 2.

10. Arendt, *Human Condition,* 7, 324. We are using the terms *Earth* and *planet* interchangeably here.

11. Arendt, *Human Condition,* 137, 152.

12. Arendt, *Human Condition,* 157.

13. Arendt, *Human Condition,* 134.

14. Arendt, *Human Condition,* 320.

15. Arendt, *Human Condition,* 322.

16. Arendt, *Human Condition,* 133.

17. Arendt, *Human Condition,* 323.

18. Hannah Arendt, *Between Past and Future: Eight Exercises in Political Thought* (1961; London: Penguin, 1977), 95.

19. Herbert Marcuse, *One-Dimensional Man: Studies in the Ideology of Advanced Industrial Society* (1964; Boston: Beacon Press, 1991), xl.

20. Marcuse, *One Dimensional Man,* xliv, 4.

21. Marcuse, *One Dimensional Man,* 7.

22. Arendt, *Human Condition,* 134.

23. Whether his philosophy was fascist and anti-Semitic to the core remains a point of intense scholarly controversy, but it is a question beyond the scope of the present discussion. See, for example, Andrew Mitchell and Peter Trawny, *Heidegger's Black Notebooks: Responses to Anti-Semitism* (New York: Columbia University Press, 2017).

24. Martin Heidegger, *The Question Concerning Technology and Other Essays* (New York: Garland Publishing, 1977), 17--19.

25. Heidegger, "Question Concerning Technology," 15.

26. Heidegger, "Question Concerning Technology," 16.

27. Martin Heidegger, "Building, Dwelling, Thinking," in *Poetry, Language, Thought,* trans. Albert Hofstadter (New York: Harper Colophon, 1971), 147.

28. Heidegger, "Building, Dwelling, Thinking," 150.

29. Heidegger, "Building, Dwelling, Thinking," 158.

30. Heidegger, "Building, Dwelling, Thinking," 159.

31. Vincent Blok, "Reconnecting with Nature in the Age of Technology," *Environmental Philosophy* 11, no. 2 (2014): 307–332.

32. Mark Stoll, "Rachel Carson's *Silent Spring,* a Book That Changed the World," Environment and Society Portal, *Virtual Exhibitions* 2012, no. 1, Rachel Carson Center for Environment and Society, version 2 (2020) at https://www .environmentandsociety.org/exhibitions/rachel-carsons-silent-spring.

33. Rachel Carson, *Silent Spring and Other Writings on the Environment,* ed. Sandra Steingraber (New York: Library of America, 2018), 9–10.

34. William Souder, *On a Farther Shore: The Life and Legacy of Rachel Carson* (New York: Crown, 2012), 7–9.

35. Carson, *Silent Spring,* 461.

36. Nancy Langston, *Toxic Bodies: Hormone Disruptors and the Legacy of DES* (New Haven: Yale University Press, 2011); Jennifer Thomson, *The Wild and the Toxic: American Environmentalism and the Politics of Health* (Chapel Hill: University of North Carolina Press, 2019).

37. Rachel Carson, *The Sea Trilogy: Under the Sea-Wind, The Sea Around Us, The Edge of the Sea* (New York: Library of America, 2021), 192.

38. Etienne Benson, *Surroundings: A History of Environments and Environmentalisms* (Chicago: University of Chicago Press, 2020). On the prehistory of environmental reflexivity, see Lydia Barnett, *After the Flood: Imagining the Global Environment in Early Modern Europe* (Baltimore: Johns Hopkins University Press, 2019); Jean Baptiste Fressoz and Fabian Locher, *Les révoltes du ciel. Une autre histoire du changement climatique XV–XVIII siècle* (Paris: Seuil, 2020).

39. Kenneth Boulding, "What Is Evolutionary Economics?" *Journal of Evolutionary Economics* 1, no. 1 (1991): 9–17, 13.

40. Kenneth Boulding, "The Economics of Space-Ship Earth," in *Collected Papers,* ed. Fred Glahe, 5 vols. (Boulder: Colorado Associated University Press, 1971–1975), 2: 389.

41. Kenneth Boulding, "Economics and Ecology," in *Collected Papers,* 3: 308.

42. Kenneth Boulding, "Economics and Ecology," 3: 309.

43. Boulding, "Is Scarcity Dead?" in *Collected Papers,* 3: 319.

44. Boulding, "Economics of Space-Ship Earth," 2: 390.

45. Boulding, "Is Scarcity Dead?" 3: 319.

46. Boulding, "Economics and Ecology," 3: 309. Boulding dated the start of science-driven economic growth to the period after 1860 and recognized a second phase of intensification after 1945. Boulding, "Is Scarcity Dead?" 3: 316.

47. Boulding, "Economics and Ecology," 3: 309.

48. Boulding, "Economics and Ecology," 3: 309.

49. Boulding, "Economics and Ecology," 3: 309.

50. Boulding, "Economics and Ecology," 3: 307; McNeill and Engelke, *Great Acceleration,* 41.

51. Boulding, "Economics and Ecology," 3: 309.

52. Boulding, "Is Scarcity Dead?" 3: 320.

53. Boulding, "Is Scarcity Dead?" 3: 320.

54. Boulding, "The Basis of Value Judgments," in *Collected Papers,* 2: 407.

55. Boulding, "The Basis of Value Judgments," 2: 410.

56. Boulding, "The Basis of Value Judgments," 2: 410.

57. Thomas Robertson, *The Malthusian Moment: Global Population Growth and the Birth of American Environmentalism* (New Brunswick, NJ: Rutgers University Press, 2012), 126, 136; quotations from Paul Sabin, *The Bet: Paul Ehrlich, Julian Simon and Our Gamble over Earth's Future* (New Haven: Yale University Press, 2014), 1, 28.

58. Donella Meadows et al., *The Limits to Growth: A Report for the Club of Rome's Project on the Predicament of Mankind* (New York: Universe Books, 1972).

59. Donald Worster, *Shrinking the Earth: The Rise and Decline of Natural Abundance* (New York: Oxford University Press, 2016), 175–179.

60. Carolyn Merchant, *The Death of Nature: Women, Ecology and the Scientific Revolution* (San Francisco: Harper Collins, 1980). See also "Interview with Carolyn Merchant for the Center for Advanced Study in the Behavioral Sciences, Stanford University, April 2021," https://ourenvironment.berkeley .edu/sites/ourenvironment.berkeley.edu/files/The%20Death%20of%20 Nature%201980%20Forty%20Years%20Later%202020.pdf.

61. Carolyn Merchant, *The Anthropocene and the Humanities: From Climate Change to a New Age of Sustainability* (New Haven: Yale University Press, 2020), 131; Val Plumwood, *Philosophy and the Mastery of Nature* (London: Routledge, 1993); Silvia Federici, *Caliban and the Witch: Women, the Body and Primitive Accumulation* (Brooklyn: Autonomedia, 2004).

62. Robert Skidelsky and Edward Skidelsky, *How Much is Enough? Money and the Good Life* (New York: Other Press, 2012).

63. Marshall Sahlins, *Stone Age Economics* (New York: Aldine de Gruyter, 1972), 1–2.

64. Sahlins, *Stone Age Economics,* 34.

65. Sahlins, *Stone Age Economics,* 39. For Sahlins' connection with Polanyi, see Gareth Dale, *Karl Polanyi: A Life on the Left* (New York: Columbia University Press, 2016), 217, 209.

66. Richard Easterlin, "Does Economic Growth Improve the Human Lot? Some Empirical Evidence," in *Nations and Households in Economics Growth: Essays in Honor of Moses Abramovitz,* ed. Paul A. David and Melvin W. Reder (New York, Academic Press, 1974), 89–125; Skidelsky and Skidelsky, *How Much is Enough?* 102–105; Milena Büchs and Max Koch, *Postgrowth and Wellbeing: Challenges to Sustainable Welfare* (London: Palgrave Macmillan, 2017).

67. Robert Leonard, "E. F. Schumacher and the Making of 'Buddhist Economics,' 1950–1973," *Journal of the History of Economic Thought* 41, no. 2 (2019): 159–186.

68. E. F. Schumacher, *Small Is Beautiful: Economics as if People Mattered* (New York: Harper and Row, 1975), 61.

69. Schumacher, *Small Is Beautiful,* 60.

70. Schumacher, *Small Is Beautiful,* 34.

71. World Commission on Environment and Development, *Our Common Future* (Oxford: Oxford University Press, 1987), 8.

72. *Our Common Future,* 43.

73. Gro Harlem Brundtland, "The Politics of Oil: A View from Norway," *Energy Policy* 16, no. 2 (1988): 102–109.

74. Brundtland, "Politics of Oil," 109.

75. "Statement of Dr. James Hansen, June 23, 1988," *Greenhouse Effect and Global Climate Change: Hearings Before the Committee on Energy and Natural Resources, United States Senate, One Hundredth Congress, First Session* (United States: US Government Printing Office, 1988), 39–41.

76. Mathis Wackernagel and William Rees, *Our Ecological Footprint: Reducing Human Impact on the Earth* (Gabriola Island, British Columbia: New Society, 1996); Paul Crutzen and Eugene Stoermer, "The Anthropocene," *IGBP Newsletter* 41 (2000), 17–18; Thomas Wiedmann and Jan Minx, "A Definition of 'Carbon Footprint,'" in *Ecological Economics: Research Trends,* ed. Carolyn C. Pertsova (New York: Nova Science, 2017), 1–11; Thomas E. Downing et al., *Social Cost of Carbon: A Closer Look at Uncertainty* (London: UK Department for Environment, Food and Rural Affairs, 2005); Rockström et al., "Planetary Boundaries."

77. Amitav Ghosh, *The Great Derangement: Climate Change and the Unthinkable* (Chicago: University of Chicago Press, 2016), 92.

Conclusion

1. Paul Falkowski, "The Power of Plankton," *Nature* 483, no. 7387 (March 1, 2012): S17–20.

2. Paul Hawken, *Drawdown: The Most Comprehensive Plan Ever Proposed to Reverse Global Warming,* 176–180; A. Mitra and S. Zaman, *Blue Carbon Reservoir of the Blue Planet* (New Delhi: Springer, 2015), 161–162.

3. Will Steffen, "Planetary Boundaries: Guiding Human Development on a Changing Planet," *Science* 347, no. 6223 (February 13, 2015).

4. Valerie Masson-Delmotte et al., *Global Warming of 1.5°C: An IPCC Special Report on the Impacts of Global Warming of 1.5°C above Pre-industrial Levels and Related Global Greenhouse Gas Emission Pathways* (Cambridge: Cambridge University Press, 2018), 261; Svetlana Jevrejeva et al., "Coastal Sea Level Rise with Warming above 2°C," *Proceedings of the National Academy of Sciences* 113, no. 47 (2016): 13342–13347; Colin Raymond, Tom Matthews, and Radley M. Horton, "The Emergence of Heat and Humidity Too Severe for Human Tolerance," *Science Advances* 6, no. 19 (2020).

5. Elizabeth Kolbert, *Under a White Sky: The Nature of the Future* (New York: Crown, 2021).

6. Diane Coyle, *The Economics of Enough: How to Run the Economy As If the Future Matters* (Princeton: Princeton University Press, 2011), 70.

7. Peter Diamandis and Steven Kotler, *Abundance: The Future Is Better than You Think* (New York: Free Press, 2012).

8. Diamandis embraces Julian Simon's critique of Paul Ehrlich and the environmentalist movement that emerged in the 1960s. See Paul Sabin, *The Bet: Paul Ehrlich, Julian Simon, and Our Gamble Over Earth's Future* (New Haven: Yale University Press, 2013).

9. Diamandis and Kotler, *Abundance,* 57–58.

10. Coyle, *Economics of Enough,* 70.

11. International Monetary Fund, *World Economic Outlook: A Long and Difficult Ascent,* October 2020.

12. Coyle, *Economics of Enough,* 69.

13. Partha Dasgupta, *The Economics of Biodiversity: The Dasgupta Review,* abridged (London: HM Treasury, 2021).

14. Dasgupta, *Economics of Biodiversity,* 10.

15. Dasgupta, *Economics of Biodiversity,* 24.

16. Dasgupta, *Economics of Biodiversity,* 80.

17. Dasgupta, *Economics of Biodiversity,* 81.

18. For an influential example of the genealogical method, see Michel Foucault, *The History of Sexuality,* vol. 1: *An Introduction,* trans. Robert Hurley (New York: Random House, 1978).

19. We take inspiration here from Quentin Skinner's defense of intellectual history as a project of excavating lost worlds in *Liberty before Liberalism* (Cambridge: Cambridge University Press, 1997), 110–112.

20. Pope Francis, *Laudato si* (Rome: FAO, 2016); Bill McKibben, *Eaarth: Making a Life on a Tough New Planet* (New York: Henry Holt, 2010); Jason Moore, *Capitalism and the Web of Life: Ecology and the Accumulation of Capital* (New York: Verso, 2015); Andreas Malm, *Fossil Capital* (New York: Verso, 2016); Holly Jean Buck, *After Geoengineering: Climate Tragedy, Repair and*

Restoration (New York: Verso, 2019); Jason Hickel, *Less Is More: How Degrowth Will Save the World* (London: William Heineman, 2020).

21. Roman Krznaric, *The Good Ancestor: A Radical Prescription for Long-Term Thinking* (New York: The Experiment, 2020), 93.

22. Compare Karen Raworth, *Doughnut Economics: Seven Ways to Think Like a 21st Century Economist* (White River Junction: Chelsea Green, 2017).

23. See, for example, Hickel, *Less Is More;* Kate Soper, *Post-Growth Living: For an Alternative Hedonism* (New York: Verso, 2020).

24. Buck, *After Geoengineering;* Max Ajl, *A People's Green New Deal* (London: Pluto Press, 2021); Troy Vettese and Drew Pendergast, *Half-Earth Socialism: A Plan to Save the Future from Extinction, Climate Change and Pandemics* (New York: Verso, 2022).

ACKNOWLEDGMENTS

Ten years ago, during a summer stroll on Södermalm in Stockholm, not far from where we both grew up, we decided to join forces and write this book together. Not long after we published our respective first monographs, both of which explore the intersection between nature and economy in early modern Europe, we felt that the time was right to write a book that explores this relationship in the *longue durée*. While one of us was focusing on the intellectual history of scarcity and the other on the intellectual history of abundance, we realized, as we were glancing out over Riddarfjärden, that we were essentially writing the same book. It was therefore a simple decision to work together. It is a decision we are very happy we made—our regular conversations over the years have sustained us intellectually and muted the howl of homesickness for two Swedes in exile.

This book is unmistakably historical, yet part of its purpose is to inform and inspire thinking about the future. Coming generations, we argue, have no choice but to develop new ways of thinking about nature and economy. As such, this is a book aimed primarily at young people across the world. In fact, we owe a profound debt to our students. The way we have organized the book and formulated its contents is strongly shaped by classroom conversations at the University of Chicago, Barnard College, and Columbia University. Students who have taken seminars with us will recognize many of the themes and ideas. This book would not have looked the same had it not been for the fresh perspectives, piercing criticisms, and intellectual creativity of our wonderful students. We thank them for making teaching a meaningful joy and for turning the classroom into an incubator for new ideas and perspectives.

One of the benefits of writing a book with a long gestation period is that we have had the opportunity to try out our ideas on many brilliant minds—professors and students alike. At a critical turning point, we were able to organize a manuscript workshop in the spring of 2021, generously funded by the Center for International Social Science Research at the University of Chicago. This experience, both sobering and rewarding, led us to rethink important features of the book. We are deeply grateful for the comments and criticisms of our friends and colleagues, including Christopher Brown, Dipesh Chakrabarty, Liz Chatterjee, Paul Cheney, Diane Coyle, Oliver Cussen, Joel Isaac, Christopher Jones, Duncan Kelly, Grant Kleiser, Charles-Francois Mathis, Ted McCormick, Tawny Paul, Steve Pincus, Lucas Pinheiro, Keith Pluymers, Adam Tooze, and Frank Trentmann. Thanks also to Carolina Marques de Mesquita for keeping extensive notes of the discussion. Many other colleagues have read parts or all of the manuscript and have generously shared their views. We thank Venus Bivar, Christophe Bonneuil, Neil Brenner, John Brewer, Dan Carey, Pierre Charbonnier, Harry Cleaver, Elizabeth Cross, Douglas Dacy, Adel Daoud, Will Derringer, Mark Fiege, Vicky de Grazia, Jo Guldi, Stephen Gross, Martha Howell, Adrian Ivashkiv, Joel Kaye, Vera Keller, Harro Maas, Stephen Macekura, Steve Medema, Arnoud Orain, Chris Otter, Gustav Peebles, Sophus Reinert, Margaret Schabas, Phil Stern, Anoush Terjanian, Troy Vettese, Deborah Valenze, Andre Wakefield, Paul Warde, Emily Webster, and Donald Worster. We are most grateful for help with technical questions about climate science and earth system science from Etienne Benson, Deborah Coen, Joshua Howe, and Emily Kern. A special thanks to Philip Rössner, John Shovlin, and Lisa Tiersten for their careful engagement and helpful suggestions on the entire manuscript. We also thank the many scholars who offered comments when we presented parts of the book at a variety of institutions, including Columbia University,

Lund University, Toulouse Institute for Advanced Studies, University of Chicago, Université de Poitiers, and the University of Cambridge.

Over the years, our research has benefited from a number of grants and fellowships. Apart from acknowledging financial assistance from our home institutions, we thank the ACLS, INET, Magnus Bergvalls Stiftelse, Sven och Dagmar Saléns Stiftelse, the Wenner-Gren Foundation, and the Institute for Advanced Study at the University of Notre Dame.

Harvard University Press has been a steadfast supporter of this project from the very start. Mike Aronson, John Kulka, and Jeff Dean offered invaluable insights and strategic advice. Emily Silk has been a truly amazing editor. Her insightful readings, astute suggestions, and infectious enthusiasm have done wonders for this book. It has been an absolute delight to work with her. We also appreciate the exceptional assistance we have received from the press's editorial team, including Julia Kirby, Sherry Gerstein, Jillian Quigley, and Stephanie Vyce. Finally, we wish to thank our two anonymous readers at Harvard, whose painstaking reviews offered welcome encouragement as well as incisive criticism.

Like any sustained academic project, ours was facilitated and supported by our respective families. The enduring love of Vicky, Monica, Langston, and Selma, each in their own ways, encouraged and inspired this book and helped put it in perspective. Everything we do has a unique and singular meaning because of them. Our final thanks go to our parents: we dedicate this book to the memory of our fathers and in celebration of our mothers.

INDEX

Page references in italics indicate a figure.